1 Chemistry of Plant Protection

Sterol Biosynthesis Inhibitors and Anti-Feeding Compounds

With Contributions by
T. Kato, W. Krämer, K.-H. Kuck,
D. M. Norris, H. Scheinpflug

With 30 Figures and 12 Tables

Springer-Verlag
Berlin Heidelberg New York Tokyo

Managing Editors:

Dr. G. Haug
Leiter des Ressorts Pflanzenschutz Anwendungstechnik
Pflanzenschutzzentrum Monheim, Bayerwerk
D-5090 Leverkusen/FRG

Prof. Dr. H. Hoffmann
Leiter des Ressorts Chemische Forschung
Pflanzenschutzzentrum Monheim, Bayerwerk
D-5090 Leverkusen/FRG

This series continues the handbook
"Chemie der Pflanzenschutz- und Schädlingsbekämpfungsmittel"
edited by R. Wegler

ISBN-3-540-13487-5 Springer-Verlag Berlin Heidelberg New York Tokyo
ISBN-0-387-13487-5 Springer-Verlag New York Heidelberg Berlin Tokyo

Typesetting and Offsetprinting: Th. Müntzer, GDR;
Bookbinding: Lüderitz & Bauer, Berlin
2154/3020-543210

Editorial Board

Editorial

The series "Chemistry of Plant Protection" continues our handbook "Chemie der Pflanzenschutz- und Schädlingsbekämpfungsmittel" edited by R. Wegler.

Aims and Scope

The development of crop protection chemistry continues apace. Important targets have been achieved, new developments are under and respect to selectivity, efficacy, mode of action environmental and toxicological compatibility as well as in the field of formulation and side-effect-research. New classes of substances showing a high effectivity even at very low dosages were discovered.

These developments have been accelerated by new biological and biochemical discoveries as well as by advances in synthetic chemistry. In order to emphasize the interaction between these and other related disciplines, the formerly distinct presentation of different fields of crop protection chemistry, like insecticide-, fungicide or herbicide research, has been abandoned. The following volumes will contain the most recent developments on new active chemical substances as well as on metabolism, residue analysis, biochemical mechanisms and other important innovations.

More or less, the volumes will be topic-oriented. Each volume will have a "volume index" which approximately characterizes the contant. The aim of the publisher and the board of editors is to produce review articles of high quality by leading scientists in the field of plant protection. Suggestions for such contributions from the plant protection community at large are of course also very welcome.

Editorial Policy

This series publishes invited, critical review articles of new developments in this area in English. The topics should comprehensively be covered and the international literature evaluated. Ideally, contributions should be from 40 to 80 typewritten pages. Experimental details, except when not published previously, should be covered by citing the appropriate references. — The responsible editor, that means, the editor who has invited the article, discusses the scope of the review with the author on the basis of a tentative outline which the author is asked to provide.

Author and editor are responsible for the scientific quality of the contribution; the editor's name appears at the end of it. — Literature should be arranged numerically, not alphabetically. An abstract of around 15 to 20 lines, and printed between author-line and table of contents, is to summarize highlights and main features of the article. — Manuscripts must be submitted, in content, language, and form satisfactory, to Springer-Verlag.

Figures and formulars should be reproducible (guide-lines are provided by Springer-Verlag). With regard to the English language, we provide the assistance of our English copy-editing department who checks and, if necessary, correct grammar and style, free of charge.

Editors and publisher make all efforts to publish the manuscripts as rapidly as possible, i.e. at the maximum 6 months after the submission of an accepted paper. This means that contributions from diverse areas of plant protection must occasionally be united in one volume. In such cases a "Volume Index" cannot meet all expectations, but will nevertheless provide more information than a mere volume number.
Each volume contains a subject index.

Editors Publisher

Table of Contents

Sterol Biosynthesis in Fungi, a Target for Broad Spectrum Fungicides

Toshiro Kato
Pesticides Research Laboratory, Takarazuka Research Center, Sumitomo Chemical Co., Ltd., 2-1, 4-Chome, Takatsukasa, Takarazuka, Hyogo 665, Japan

Ergosterol, a principal sterol in most fungi, is an indispensable component in the membrane structures. A recent innovation in fungicide research is the findings of fungicides which inhibt ergosterol biosynthesis in fungi. They block the biosynthetic reactions in the conversion of lanosterol to ergosterol. Inhibitors of 14α-demethylation of lanosterol and 24-methylenedihydrolanosterol are the most widely discovered segment of sterol biosynthesis inhibitors; they include substituted pyrimidines, pyridines, piperazines, triazoles and imidazoles. Morpholine fungicides are the second segment of the inhibitors; they inhibit $\triangle^8 \rightarrow \triangle^7$ isomerization (the conversion of fecosterol to episterol). Other inhibitors block $\triangle^{14}$ reduction, $\triangle^{22}$ unsaturation, C-24 transmethylation and $\triangle^{24(28)}$ reduction.

The effects of the inhibitors on membrane permeability and fungal morphology are also described.

1 Introduction

Living organisms are separated from the outside world by their membranes, which perform important roles for maintaining their lives. Sterols are indispensable components in eukaryotic organisms for stabilizing membranes composed of phospholipid bilayers. Most prokaryotic organisms, on the other hand, do not contain sterols in their membranes, although recent studies indicated that the bacterium *Methylococcus capsulatus* produces sterols [1] and many strains of *Mycoplasma*, bacteria without cell walls, require sterols as an essential growth factor [2]. Some bacterial species synthesize hopanoids which are hypothesized to function like sterols in eukaryotic organisms [3].

Significance of ergosterol in fungi has been recognized from the fact that the yeast *Saccharomyces cerevisiae* requires ergosterol for growth, when cultured in a strictly anaerobic condition [4]. Electron microscopic studies showed that membrane structures disappeared in the yeast grown anaerobically without a supplement of ergosterol [5,6], indicating the necessity of ergosterol for membrane biogenesis. Further confirmation of the significant role of ergosterol in the membranes was derived from the studies on polyene antibiotics. In the presence of polyenes such as nystatin, filipin and amphotericin B, fungal cells lose selective permeability of the membranes, since polyenes interact with ergosterol, such that the structure of the fungal membranes is disrupted [7].

During the last decade, a number of inhibitors of ergosterol biosynthesis in fungi have been found as fungicides for agricultural or clinical uses. Extensive studies have been made to elucidate the mode of action of these fungicides and to determine the exact site of inhibition during ergosterol biosynthesis. This chapter describes these fungicides with a special attention to the biosynthetic reactions inhibited by them.

2 Composition of Sterols in Fungi

It is generally accepted that fungi possess ergosterol as a principal sterol, whereas major sterols in mammals and plants are cholesterol and sitosterol or stigmasterol, respectively. Most species of the Ascomycetes, the Basidiomycetes and the Fungi Imperfecti produce ergosterol as an abundant sterol [8,9]. In rust fungi, however, stigmast-7-enol and stigmasta-7,24(28)-dienol are predominant sterols [10,11]. Furthermore, some species of the Phycomycetes fungi produce other sterols rather than ergosterol. Species of the Saprolegniales and the Leptomitales contain cholesterol, 24-methylene cholesterol (ergosta-5,24(28)-dienol) or fucosterol (stigmasta-5,24(28)-dienol) as a dominant sterol, whereas species of the Mucorales possess ergosterol or 22-dihydroergosterol (ergosta-5,7-dienol) [12].

The Pythiaceous fungi, *Pythium* and *Phytophthora*, on the other hand, do not synthesize sterols and require an exogenous supply of a suitable sterol for sexual or zoosporangial reproduction [13], although one of the Pythiaceous fungi, *Zoophagus insidians* was recently reported to synthesize ergosterol[14].

Most fungi usually contain various sterols in small quantities in addition to ergosterol or other predominant sterols. These include lanosterol, 24-methylenedihydrolanosterol (24-methylene-lanost-8-enol), zymosterol (cholesta-8,24-dienol), fecosterol (ergosta-8,24(28)-dienol), episterol (ergosta-7,24(28)-dienol), fungisterol (ergost-7-enol) and others, which are intermediates in ergosterol biosynthesis. In contrast, polyene antibiotic-resistant mutants of *S. cerevisiae* [15–18], *Candida albicans* [19,20], *Aspergillus fennelliae* [21], *Neurospora crassa* [22], *Ustilago maydis* [23] and *Aspergillus nidulans* [24] accumulate these intermediates or other unusual sterols instead of ergosterol, so that they can survive in the presence of the toxicants.

3 Biosynthesis of Ergosterol in Fungi

Sterols in all living organisms are biosynthesized from acetic acid via mevalonic acid and squalene as key intermediates [25,26], although the carbon atoms at C-24 originate from methionine [27]. Lanosterol is the first cyclization product in mammals and fungi, while plants produce cycloartenol instead of lanosterol [27]. The conversion of lanosterol to ergosterol occurs through several alternative pathways in yeasts and filamentous fungi [28,29] and the major sequences of transformation are depicted in Fig. 1. Biosynthetic reactions leading to ergosterol include; (i) methylation at C-24, (ii) demethylation of the three extra methyl groups at C-4 and C-14, (iii) isomerization of the Δ^8 double bond to Δ^7, (iv) introduction of a Δ^5 double bond, (v) reduction of the $\Delta^{24(28)}$ double bond and (vi) introduction of a Δ^{22} double bond. During these reactions, a major difference of the biosynthetic sequence between filamentous fungi and yeasts is the order of transmethylation. Since filamentous fungi contain 24-methylenedihydrolanosterol [30–32], it appears that transmethylation occurs at the lanosterol level. *S. cerevisiae*, however, does not produce this sterol in normal conditions and the methylation at C-24 is believed to occur after all of the extra methyl groups are eliminated [33,34]. The yeasts, *Candida utilis* and *C. albicans* contain a small amount of 24-methylenedihydrolanosterol, implicating that transmethylation occurs at the lanosterol level [35] but this sequence does not appear to predominate [36]. It is generally accepted that S-adenosyl-methionine serves as the methyl donor for C-24 alkylation in ergosterol biosynthesis [37].

Demethylation procedes at C-14 in both yeasts and filamentous fungi; the methyl group is converted successively to hydroxymethyl and formyl intermediates, and finally removed as formic acid [38]. The involvement of cytochrome P-450 in this oxidative elimination is well documented [39–44]. The immediate product after demethylation is a $\Delta^{8,14}$ sterol, of which the Δ^{14} double bond is then reduced to give the corresponding Δ^8 sterol. Subsequently, the 4,4-dimethyl groups are removed as carbon dioxide by stepwise oxidation through the formation of hydroxymethyl, formyl and carboxyl intermediates [26].

Fecosterol, a key intermediate after transmethylation and demethylation, is further converted to episterol by isomerization of the Δ^8 double bond to Δ^7. The following reaction sequence predominating in yeasts and filamentous fungi is not fully established, but introduction of double bonds at C-5 and at C-22 positions may precede reduction of the $\Delta^{24(28)}$ double bond [45,46]. Since the enzymes involved in the conversion of lanosterol to ergosterol possess relatively poor substrate specificity, the reaction sequences described here are not always true and may vary depending on the strains or fungal species.

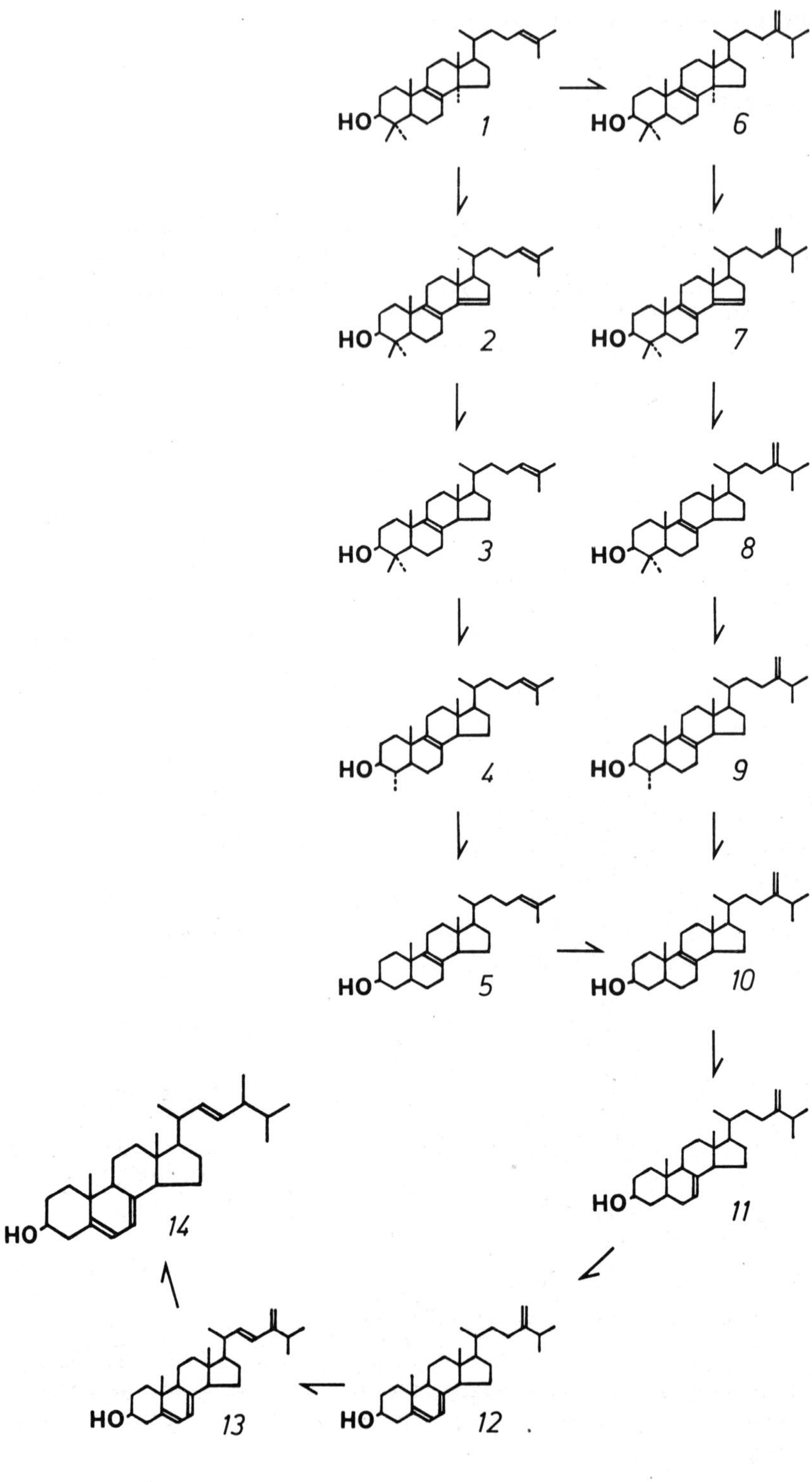
HO
1
HO
6
HO
2
HO
7
HO
3
HO
8
HO
4
HO
9
HO
5
HO
10
HO
11
HO
14
HO
13
HO
12

4 Inhibitors of Ergosterol Biosynthesis

4.1 14α-Demethylation Inhibitors

4.1.1 Pyrimidines and Pyridines

Triarimol (Fig. 2) is a pyrimidine fungicide effective in controlling apple scab [47] and powdery mildews [48]. Since the fungicide is structurally similar to triparanol (Fig. 5), an inhibitor of cholesterol biosynthesis in mammalian systems, Ragsdale and Sisler [49] tested the effect of triarimol on sterol synthesis in *U. maydis* and found a strong inhibition of ergosterol formation. The fungicide did not primarily affect other metabolic activities such as oxidation of glucose or acetate, and DNA, RNA and protein syntheses [50]. In the treated culture of *U. maydis*, 24-methylenedihydrolanosterol, obtusifoliol (4α,14α-dimethyl-ergosta-8,24(28)-dienol) and 14α-methyl-ergosta-8,24(28)-dienol (Fig. 3) accumulated [51]. Since all of these accumulated sterols possess extra methyl group(s) at C-4 and C-14, it was thought that the fungicide blocked demethylation reactions. By employing a cell-free system of *S. cerevisae*, however, it became apparent that triarimol selectively inhibited 14α-demethylation [52]. The presence of the C-4 methyl group(s) in the accumulated sterols therefore indicates that demethylation at C-4 occurs at a greatly reduced rate, when sterols possess the 14α-methyl group. Mitropoulos and Gibbons [53] also found that triarimol markedly prevented the incorporation of radioactivity from mevalonate-2-^{14}C into cholesterol in rat liver homogenates. When 24,25-dihydrolanosterol (lanost-8-enol)-32-^{14}C was incubated with rat liver microsomes in the presence of the fungicide, the release of ^{14}C-formic acid was suppressed [53], confirming the inhibition of 14α-demethylation. Triarimol is also toxic to the growth of *Chlorella* species and the treatment of the fungicide resulted in the accumulation of 14α-methyl sterols, implicating that 14α-demethylation is also arrested in the biosynthesis of plant sterols [54].

Fenarimol and nuarimol (Fig. 2) are close structural relatives of triarimol. Both fungicides interrupt ergosterol biosynthesis in *Ustilago avenae* in a manner similar to triarimol [55]. Similarly, in bramble suspension cultures, a large accumulation of 14α-methyl sterols occurred when treated with fenarimol [56].

Buthiobate (Fig. 2), a pyridine fungicide for controlling powdery mildew of cucurbits [57], interferes with ergosterol biosynthesis in *Monilinia fructigena* with a concomitant accumulation of 14α-methyl sterols [32, 58]. In cell-free homogenates of *S. cerevisae*, the fungicide blocked 14α-demethylation of lanosterol [52]. Aoyama et al. [59] recently showed that buthiobate inhibited 14α-demethylation of lanosterol in the reconstituted enzyme system of cytochrome P-450 and NADPH-cytochrome P-450 reductase purified from *S. cerevisiae*. Induction of a Type II spectral change of

---

◀ **Fig. 1.** Biosynthetic sequences from lanosterol to ergosterol.
1, Lanosterol; 2, 4,4-Dimethyl-cholesta-8,14,24-trienol; 3, 4,4-Dimethyl-cholesta-8,24-dienol; 4, 4α-Methylcholesta-8,24-dienol; 5, Zymosterol; 6, 24-Methylenedihydrolanosterol; 7, 4,4-Dimethyl-ergosta-8,14,24(28)-trienol; 8, 4,4-Dimethyl-ergosta-8,24(28)-dienol; 9, 4α-Methyl-ergosta-8,24(28)-dienol; 10, Fecosterol; 11, Episterol; 12, Ergosta-5,7,24(28)-trienol; 13, Ergosta-5,7,22,24(28)-tetraenol; 14, Ergosterol

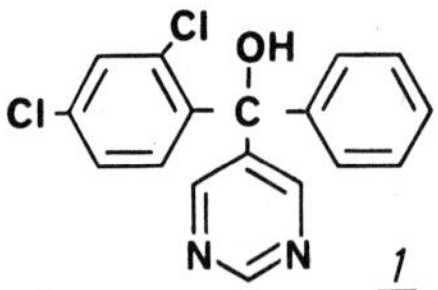

1

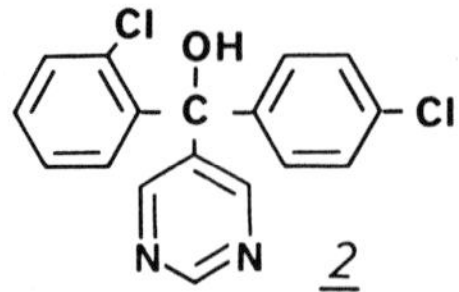

2

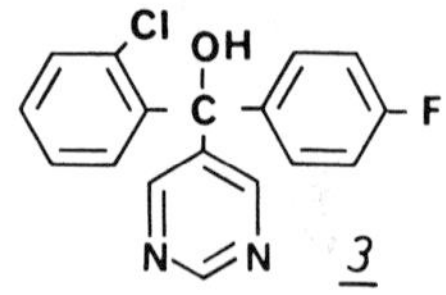

3

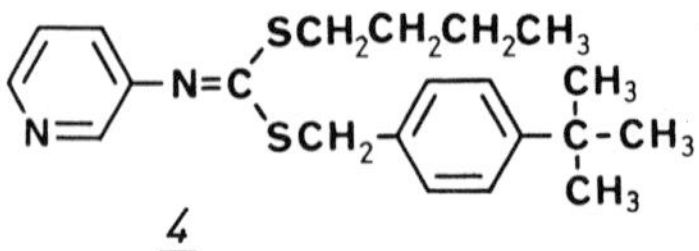

4

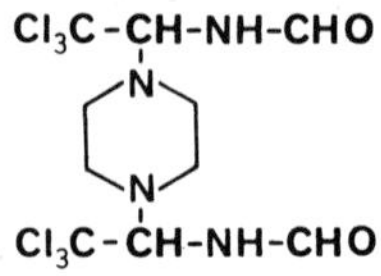

5

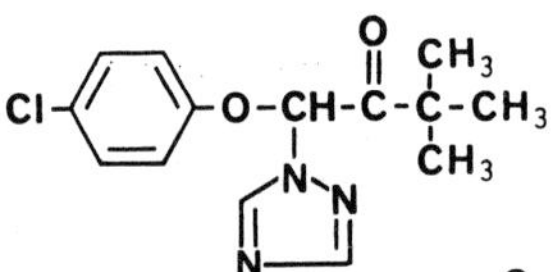

6

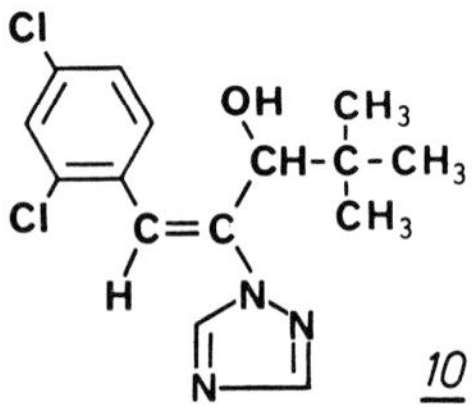

10

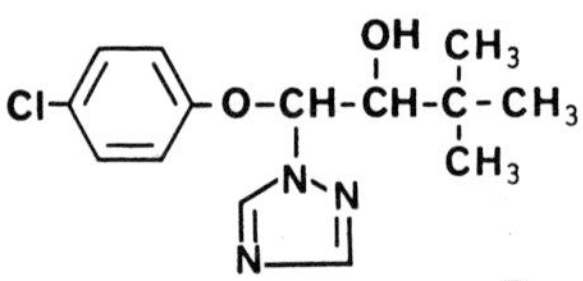

7

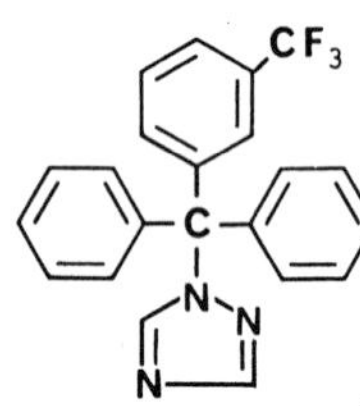

11

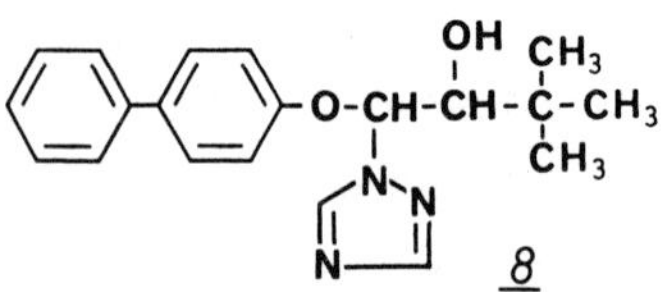

8

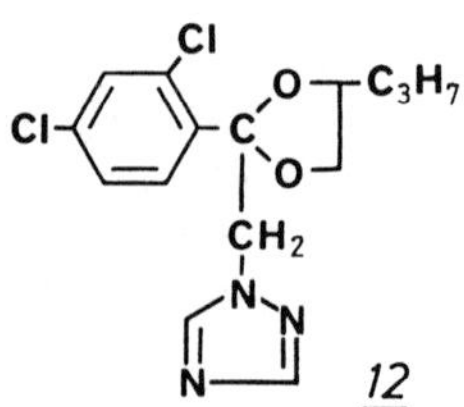

12

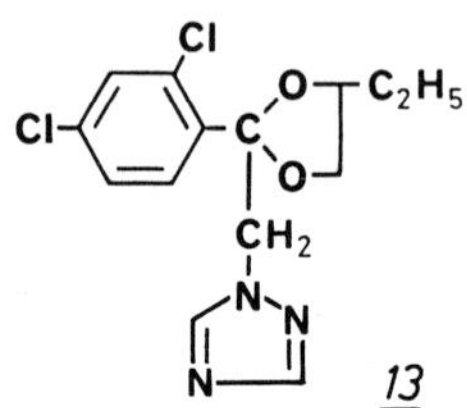

13

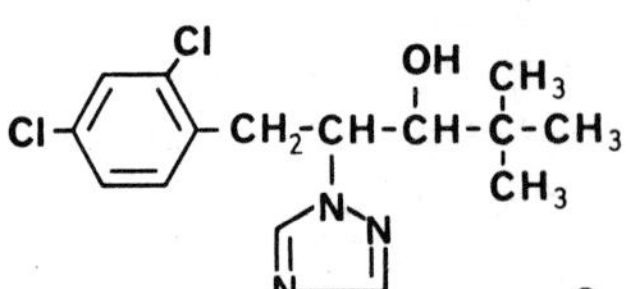

9

Fig. 2. Chemical structures of 14α-demethylation inhibitors.
1, Triarimol; *2*, Fenarimol; *3*, Nuarimol; *4*, Buthiobate; *5*, Triforine; *6*, Triadimefon; *7*, Triadimenol; *8*, Bitertanol; *9*, Diclobutrazol; *10*, S-3308; *11*, Fluotrimazole; *12*, Propiconazole; *13*, Etaconazole

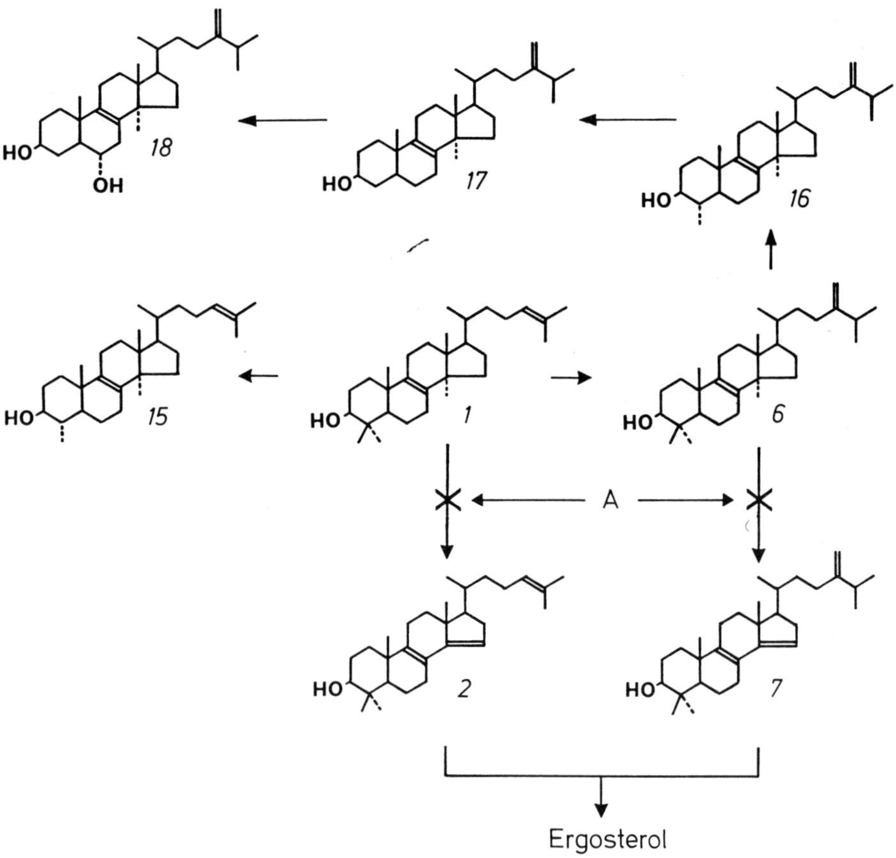

Fig. 3. Biosynthetic reactions blocked by 14α-demethylation inhibitors (A) and resultant altered pathways.
1, Lanosterol; 2, 4,4-Dimethyl-cholesta-8,14,24-trienol; 6, 24-Methylenedihydrolanosterol; 7, 4,4-Dimethyl-ergosta-8,14,24(28)-trienol; 15, 4α,14α-Dimethyl-cholesta-8,24-dienol; 16, Obtusifoliol; 17, 14α-Methyl-ergosta-8,24(28)-dienol; 18, 14α-Methyl-ergosta-8,24(28)-dien-3β,6α-diol

the cytochrome by buthiobate suggests that the fungicide binds to the cytochrome and occupies the active site to inhibit the catalytic activity. Since 14α-demethylation occurs through the formation of hydroxymethyl and formyl intermediates [38] and no such intermediates were detected in the inhibited cultures, buthiobate is considered to block the initial hydroxylation by the cytochrome P-450-involved enzyme system.

4.1.2 Piperazines

Triforine (Fig. 2) is a piperazine fungicide effective in controlling powdery mildews, rusts and apple scab [60–63]. Sherald et al. [64] pointed out similarities in the antifungal mode of action between triforine and triarimol. Both compounds caused similar morphological changes in *Cladosporium cucumerinum* and triarimol-resistant mutants of this fungus showed cross resistance to triforine. The fungicide inhibited ergosterol biosynthesis in *N. crassa* [64] and in *Aspergillus fumigatus* [65], resulting in the accumulation of 14α-methyl sterols.

4.1.3 Triazoles

Triadimefon (Fig. 2) is a triazole fungicide for controlling powdery mildews and rusts [66]. The inhibition of ergosterol biosynthesis was noted in the treated cultures of *U. avenae* [67] and *U. maydis* [68] without significant effects on respiration and syntheses of nucleic acids and protein [69]. Although the exact identification of the sterols was not conducted, 4,4-dimethyl and 4α-methyl sterols, which seemed to possess the 14α-methyl group, accumulated in the treated cultures. When rat liver homogenates were incubated with mevalonate-2-^{14}C in the presence of triadimefon, a large accumulation of radioactivity was found in the fraction of 4,4-dimethyl sterols, suggesting a strong inhibition of 14α-demethylation [70]. A similar result was obtained by employing a cell-free system of *S. cerevisiae* [70]. Although a major effect of triadimefon on sterol synthesis is regarded as the inhibition of 14α-demethylation, it was reported that a "feed-back" inhibition of 3-hydroxy-3-methylglutaryl (HMG)-CoA-reductase occurred by the accumulation of lanosterol in *S. cerevisiae* [71].

Fluotrimazole [72] (Fig. 2) prevents sporidial multiplication of *U. avenae* and inhibits ergosterol synthesis with a concomitant accumulation of 4,4-dimethyl and 4α-methyl sterols [67, 73]. Glucose oxidation and syntheses of RNA, DNA and protein were not prominently affected during the early incubation period with the fungicide [73]. The results therefore suggested that fluotrimazole is a specific inhibitor of ergosterol biosynthesis probably at the site of 14α-demethylation.

Two related fungicides, etaconazole and propiconazole (Fig. 2) have been introduced as the fungicides for controlling foliar diseases of cereals and fruit crops [74]. The former compound apparently inhibited the synthesis of ergosterol in *U. maydis* with a concomitant accumulation of 14α-methyl sterols [75, 76]. During a long incubation period, another sterol metabolite, 14α-methyl-ergosta-8,24(28)-dien-3β,6α-diol, appeared in the etaconazole-treated culture of *U. maydis*, which was probably metabolized from 14α-methyl-ergosta-8,24(28)-dienol [76] (Fig. 3). The latter fungicide, propiconazole, likewise prevented sterol synthesis in *Taphrina deformans* [70], the principal sterol of which is brassicasterol (ergosta-5,22-deinol) [77]. Since lanosterol and 24-methylenedihydrolanosterol accumulated to a large extent in the treated culture, the fungicide was considered to be an inhibitor of 14α-demethylation.

A triazole fungicide S-3308 (Fig. 2) is being developed as the fungicide for controlling powdery mildews and rusts of cereals and fruit crops, and other plant diseases. In *U. maydis* culture, the fungicide produced swollen and multicellular sporidia, and inhibited sterol synthesis with little effect on the respiration and the syntheses of nucleic acids and protein at low fungitoxic concentrations [78]. A large accumulation of 14α-methyl sterols in the treated culture suggested the inhibition of 14α-demethylation. A closely related fungicide diclobutrazol [79] (Fig. 2) inhibited the incorporation of radioactivity from mevalonate-2-^{14}C into 4-desmethyl sterols in a cell-free system of *S. cerevisiae* [70]. A concomitant accumulation of radioactivity in the fraction of 4,4-dimethyl sterols implicated the inhibition of 14α-demethylation.

Bitertanol (Fig. 2) is a fungicide structurally related to triadimefon. Ergosterol synthesis in *U. avenae* was shown to be arrested by this fungicide, and 4,4-dimethyl and 4α-methyl sterols accumulated in the culture [80]. Triadimenol, a reduced product of triadimefon [81,82] (Fig. 2), is a fungicide for controlling seed-borne pathogens [83] and it is assumed from the chemical structure to inhibit 14α-demethylation.

Other triazole fungicides include flutriafol (pp450) [83-1] and penconazole (CGA 71818) [83-2], experimental agricultural fungicides for cereals and fruit crops, respectively, and oriconazole (itraconazole) [83-3] and vibunazole (BAY n7133) [83-4], antimycotics for clinical uses.

4.1.4 Imidazoles

Imazalil (Fig. 4) is an imidazole fungicide toxic particularly to the species of *Penicillium* [84]. It prevented ergosterol synthesis in *U. avenae* [85], *U. maydis* [86], *B. cinerea* [86], *Penicillium expansum* [86,87] and *A. nidulans* [88], accumulating 4,4-dimethyl

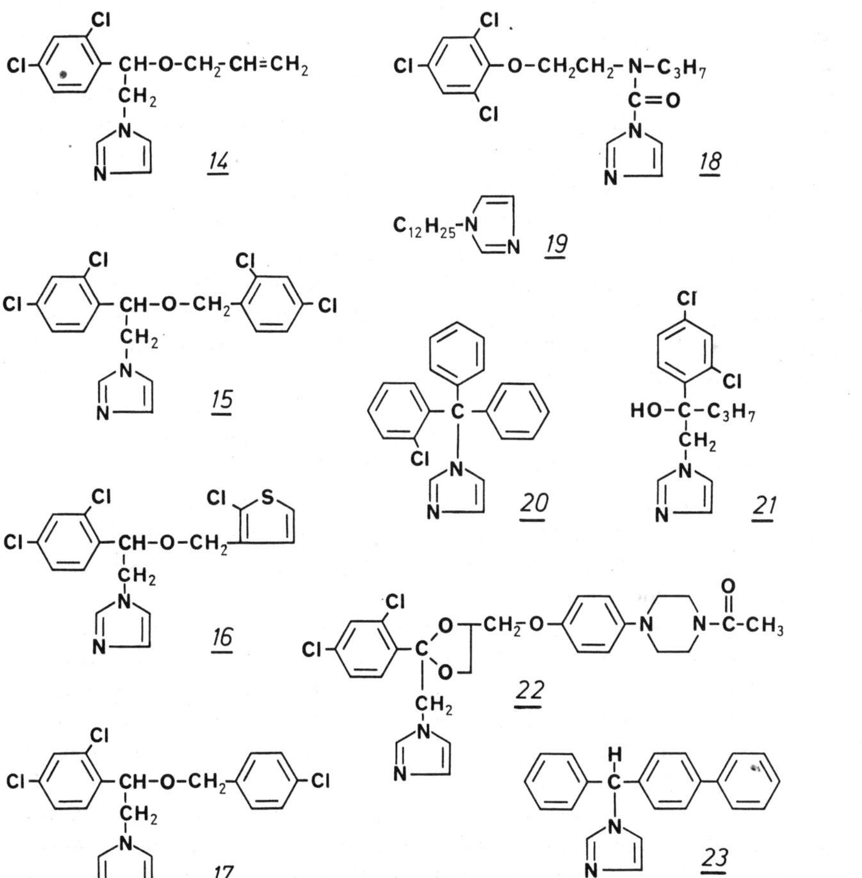

Fig. 4. Chemical structures of 14α-demethylation inhibitors.
14, Imazalil; *15*, Miconazole; *16*, Tioconazole; *17*, Econazole; *18*, Prochloraz; *19*, N-Dodecyl-imidazole; *20*, Clotrimazole; *21*, UK-38,667; *22*, Ketoconazole; *23*, Bifonazole

and 4α-methyl sterols in the culture. Fungal respiration and syntheses of macromolecules did not appear to be inhibited during an early incubation period, suggesting that the fungicide primarily affects sterol synthesis [85,88]. Prochloraz [89] (Fig. 4), a broad spectrum fungicide for foliar diseases, was also indicated to be a potent inhibitor of 14α-demethylation in a yeast cell-free system [70].

Several antifungal imidazole compounds for clinical uses are known to inhibit the biosynthesis of ergosterol in fungal systems. Clotrimazole [90] (Fig. 4) suppressed sporidial multiplication of *U. avenae* and produced swollen and branched sporidia [73]. Ergosterol formation was apparently reduced in the treated cultures of *U. avenae* [73] and *S. cerevisiae* [91] with accumulation of 4,4-dimethyl and 4α-methyl sterols. Miconazole [92] (Fig. 4) inhibited ergosterol biosynthesis in *C. albicans* and accumulated 24-methylenedihydrolanosterol, lanosterol, obtusifoliol, 4α, 14α-dimethyl-cholesta-8,24-diol and 14α-methyl-ergosta-8,24(28)-dienol in the treated culture [93] (Fig. 3). A similar observation was made in *U. maydis* [94] and in *S. cerevisiae* [91]. Ketoconazole [95] (Fig. 4) affected the biosynthesis of ergosterol in *C. albicans* in a manner quite similar to miconazole [96]. In cell-free homogenates of *C. albicans*, the conversion of mevalonate-2-^{14}C to 4-desmethyl sterols was strongly suppressed by clotrimazole, miconazole and ketoconazole, accumulating radioactivity in the 4,4-dimethyl and 4α-methyl sterol fractions [97]. Similar results were obtained with other antifungal drugs, tioconazole [98] and econazole [99] (Fig. 4) by employing the same cell-free system of *C. albicans* [97].

An experimental antifungal drug, UK-38,667 (Fig. 4) was also described to be an inhibitor of 14α-demethylation in whole cells and cell-free homogenates of *C. albicans* [100]. Furthermore, a fungitoxic 1-alkyl-imidazole, N-dodecyl-imidazole (Fig. 4) was reported to inhibit 14α-demethylation in *U. maydis* [94].

Other imidazole antimycotics, lombazole and climbazole [100-1], oxiconazole [100-2], fenticonazole [100-3] and sulconazole [100-4], are also supposed to inhibit ergosterol biosynthesis by the same mechanism.

4.2 $\Delta^8 \rightarrow \Delta^7$ Isomerization Inhibitors

Tridemorph (Fig. 5) is a systemic fungicide for controlling powdery mildews of cereals. Although at a high concentration (1×10^{-4} M) the fungicide prevented spore germination of *B. cinerea* at a low concentration (1×10^{-5} M) it caused morphological abnormality similar to that by buthiobate [101]. In the same organism, ergosterol synthesis was arrested by the fungicide and fecosterol, ergosta-8,22,24(28)-trienol and ergosta-8,22-dienol accumulated in this order in the treated culture [101,102] (Fig. 6). All the accumulated sterols were Δ^8 sterols, suggesting that tridemorph inhibits $\Delta^8 \rightarrow \Delta^7$ isomerization. The accumulation of fecosterol was also found in the tridemorph-treated culture of *S. cerevisiae* [102]. The Oomycete *Saprolegnia ferax* normally produces cholesterol, fucosterol and 24-methylene cholesterol [103]. The tridemorph-treated culture of this fungus accumulated Δ^8 sterols including zymosterol, fecosterol and stigma-8,E-24(28)-dienol [104].

In plant cells, however, cycloeucalenol-obtusifoliol isomerization appears to be prevented by tridemorph [105,106], but low concentrations of the fungicide caused accumulation of Δ^8 sterols in maize seedling, implicating the inhibition of $\Delta^8 \rightarrow \Delta^7$ isomerization in plant cells [106].

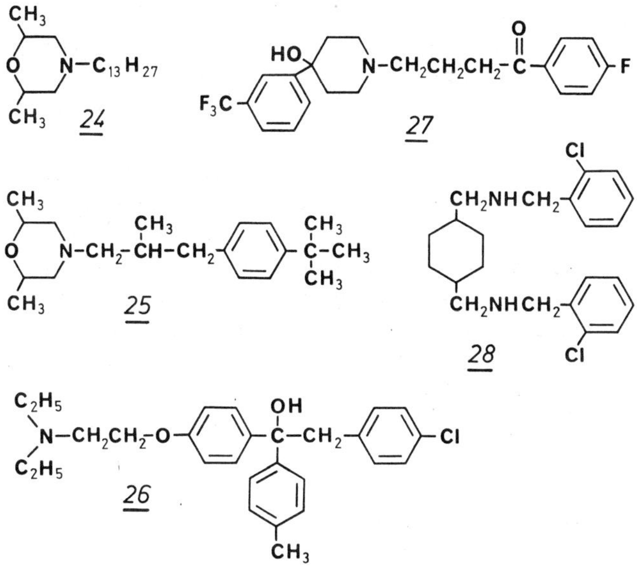

Fig. 5. Chemical structures of $\triangle^8 \rightarrow \triangle^7$ isomerization inhibitors.
24, Tridemorph; *25*, Fenpropemorph; *26*, Triparanol, *27*, Trifluperidol; *28*, AY-9944

In contrast, Kerkenaar et al. [107] reported that the accumulated sterol in the tridemorph-treated culture of *U. maydis* was ignosterol (ergosta-8,14-dienol) (Fig. 8). They thought therefore that in *U. maydis* the fungicide blocks $\triangle^{14}$ reduction.

The hypocholesteremic drug trifluperidol [108] (Fig. 5) is toxic to the growth of *S. cerevisiae* and inhibits ergosterol synthesis. The treated culture contained large quantities of $\triangle^8$ sterols including ergosta-8,24(28-dienol, ergost-8-enol and ergosta-8,22-dienol [109] (Fig. 6). Thus, the major block in the sterol biosynthetic pathway appears to be $\triangle^8 \rightarrow \triangle^7$ isomerization in this organism. Triparanol (Fig. 5), a hypocholesteremic agent in experimental animals [110], has been shown to block the conversion of desmosterol (24-dehydrocholesterol) in mammals [111]. The reduction of the $\triangle^{24(25)}$ double bond is likely to be impaired by the drug. In *S. cerevisiae*, however, triparanol was shown to inhibit $\triangle^8 \rightarrow \triangle^7$ isomerization, since less amount of ergosterol and larger amounts of $\triangle^8$ sterols, ergosta-8,24(28)-dienol, ergost-8-enol and ergosta-5,8,22-trienol were found in the treated culture [112] (Fig. 6).

Another hypocholesteremic drug AY-9944 (Fig. 5) inhibits cholesterol synthesis by blocking the conversion of 7-dehydrocholesterol to cholesterol in mammalian systems [113]. Toxicity of the drug to *S. cerevisiae* is not prominent. But the composition of sterols was markedly altered especially in the sterol ester fraction where a $\triangle^8$ sterol, fecosterol, was found at a highly increased level and ergosterol content decreased, showing a possible inhibition of $\triangle^8 \rightarrow \triangle^7$ isomerization [114]. A similar change of the sterol composition occurred to a lesser extent in the free sterol fraction. This is possibly because ergosterol tends to predominate in the sterol fraction, whereas the sterol intermediates are esterified to a greater extent [115].

Bramble cells produce sitosterol and other $\triangle^5$ sterols as principal sterols. In the AY-9944-treated culture, however, only a small amount of $\triangle^5$ sterols was detected

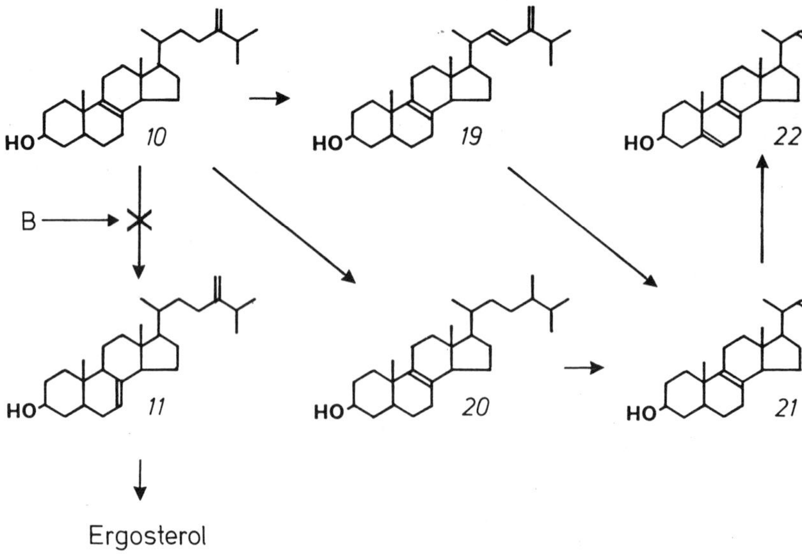

Fig. 6. Biosynthetic reaction blocked by $\triangle^8 \rightarrow \triangle^7$ isomerization inhibitors (B) and resultant altered pathways.
10, Fecosterol; 11, Episterol; 19, Ergosta-8,22,24(28)-trienol; 20, Ergost-8-enol; 21, Ergosta-8,22-dienol; 22, Ergosta-5,8,22-trienol

and $\triangle^8$ sterols including 24-ethyl-cholest-8-enol, stigmasta-8,Z-24(28)-dienol greatly accumulated [116]. This result is consistent with the effect on *S. cerevisiae.*

A recently introduced fungicide fenpropemorph [117] (Fig. 5) is likely to be an inhibitor of $\triangle^8 \rightarrow \triangle^7$ isomerization, because of the structural similarity to tridemorph.

4.3. $\triangle^{14}$ Reduction Inhibitors

The antimycotic azasterol A25822B (Fig. 7) is a mold metabolite isolated from *Geotrichum flavo-brunneum* and structurally similar to fecosterol [118]. In very low concentrations (1.0 to 10.0 ng/ml), the azasterol appeared to inhibit $\triangle^{14}$ reduction, resulting in accumulation of ignosterol [119,120] (Fig. 8). The $\triangle^{14}$ reductase activity was inhibited in the crude enzyme preparation of *S. cerevisiae* by the antimycotic [121]. The specific inhibition of $\triangle^{14}$ reduction was also recognized in the phytopathogenic fungus *U. maydis* [122]. If one considers the normal pathway of ergosterol biosynthesis, the inhibition of $\triangle^{14}$ reductase might be expected to result in an immediate accumulation of 4,4-dimethyl-cholesta-8,14,24-trienol in *S. cerevisiae* or 4,4-dimethyl-ergosta-8,14,24(28)-trienol in filamentous fungi (Fig. 8). A rapid accumulation of ignosterol in the treated culture therefore indicates that the presence of the $\triangle^{14}$ double bond does not hurt transmethylation at C-24, demethylation at C-4 and $\triangle^{24(28)}$ reduction. The $\triangle^{14}$ reductase is also probably inhibited in azasterol-treated cultures of bramble cells, in which sitosterol and other $\triangle^5$ sterols almost disappeared, whereas $\triangle^{8,14}$ sterols accumulated in large quantities [123].

AY-9944, which was supposed to be an inhibitor of the $\triangle^8 \rightarrow \triangle^7$ isomerase in *S. cerevisiae* [114], caused a rapid accumulation of $\triangle^{8,14}$ sterols and $\triangle^8$ sterols in

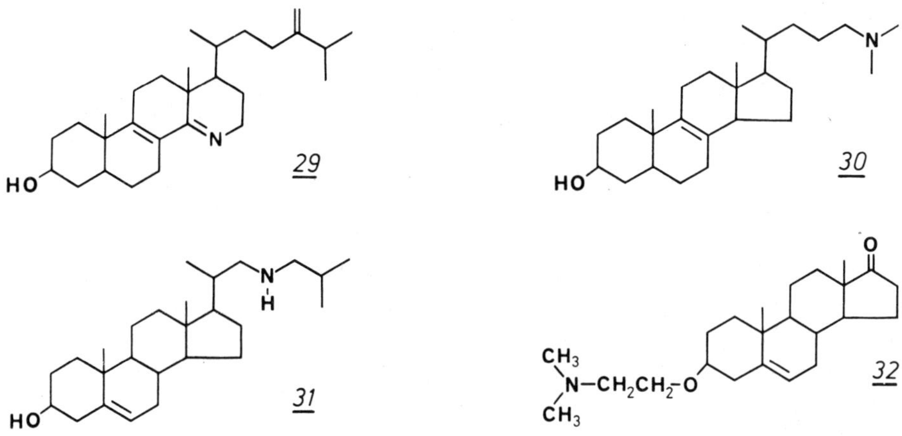

Fig. 7. Chemical structures of inhibitors of Δ^{14} reduction and side chain modifications.
29, Azasterol A25822B; *30*, 25-Aza-24,25-dihydrozymosterol; *31*, 23-Azacholesterol; *32*, DMAE-DHA

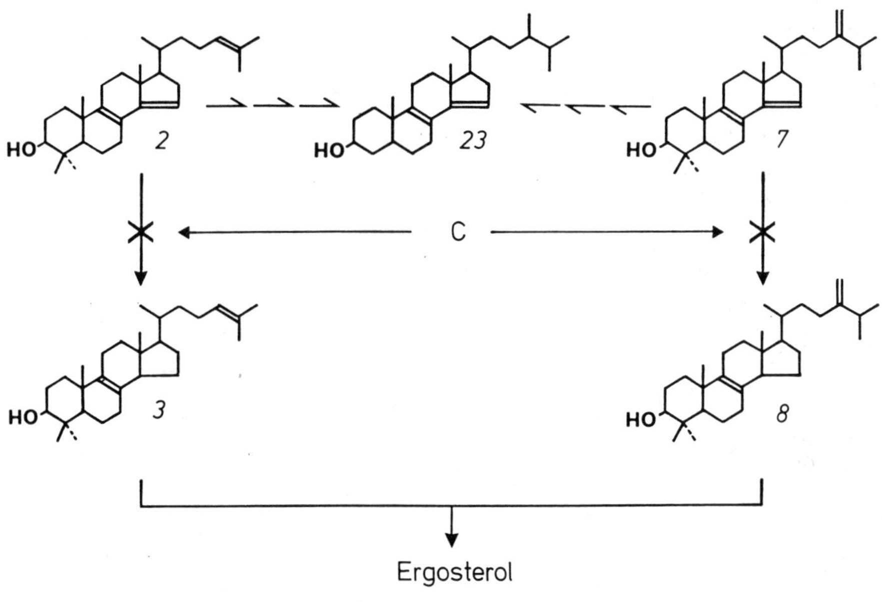

Fig. 8. Biosynthetic reactions blocked by Δ^{14} reduction inhibitors (C) and resultant altered pathways. 2, 4,4-Dimethyl-cholesta-8,14,24-trienol; 3, 4,4-Dimethylcholesta-8,24-dienol; 7, 4,4-Dimethyl-ergosta-8,14,24(28)-trienol; 8, 4,4-Dimethyl-ergosta-8,24(28)-dienol; 23, Ignosterol

Chlorella ellipsoidea, which normally produces only Δ^5 sterols [124,125]. This indicates that the drug probably interferes with both Δ^{14} reductase and $\Delta^8 \rightarrow \Delta^7$ isomerase. The inhibition of Δ^{14} reductase by AY-9944 was similarly noted in rat liver homogenates at a high concentration (1×10^{-4} M) [126,127], although at a low concentration (1×10^{-5} M) it prevented the conversion of 7-dehydrocholesterol to cholesterol [128].

4.4 Inhibitors of Side Chain Modifications

In the conversion of lanosterol to ergosterol, the following biosynthetic reactions occur for the modification of the side chain; (1) transmethylation at C-24, (2) reduction of the 24(28) double bond and (3) introduction of a 22(23) double bond.

Avruch et al. [129] synthesized 25-aza-24,25-dihydrozymosterol (Fig. 7), expecting it to be an inhibitor of C-24 methyltransferase. In fact, they found that the azasterol-treated culture of *S. cerevisiae* did not produce ergosterol and accumulated zymosterol and cholesta-5,7,22,24-tetraenol, indicating blockage of the side chain methylation (Fig. 9). 23-Azacholesterol (Fig. 7), a synthetic azasterol toxic to *C. albicans* and *S. cerevisiae* [130], inhibited $\Delta^{24(28)}$ reductase at low concentrations (2×10^{-7} to 1×10^{-6} M) in *S. cerevisiae*, since ergosta-5,7,22,24(28)-tetraenol, an immediate precursor of ergosterol, accumulated in the culture [131] (Fig. 10). At higher concentrations (1×10^{-5} M), the compound appeared to inhibit C-24 methyltransferase [131]. Other synthetic azasterols, 25-azacholesterol and 25-azacholestanol likewise blocked the C-24 methyltransferase at 1×10^{-6} M in *S. cerevisiae* [132]. Although a naturally occurring azasterol A25822B inhibited Δ^{14} reductase at low concentrations in *S. cerevisiae*, the higher concentrations blocked transmethylation [133] and $\Delta^{24(28)}$ reduction [120] in enzyme preparations of the yeast.

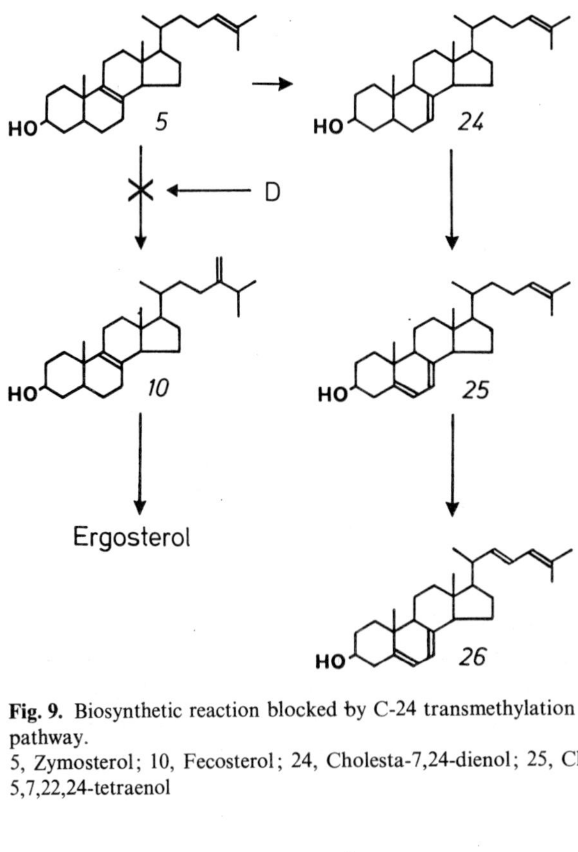

Fig. 9. Biosynthetic reaction blocked by C-24 transmethylation inhibitors (D) and resultant altered pathway.
5, Zymosterol; 10, Fecosterol; 24, Cholesta-7,24-dienol; 25, Cholesta-5,7,24-trienol; 26, Cholesta-5,7,22,24-tetraenol

The hypocholesteremic compound DMAE-DHA [134] (Fig. 7) inhibits the cyclization of 2,3-oxidosqualene to lanosterol in *S. cerevisiae* at 1×10^{-4} M [135,136]. With a prolonged incubation, accumulated 2,3-oxidosqualene in the treated culture was converted to an unusual metabolite, 2,3;22,23-dioxidosqualene which became the major product in the sterol metabolism of the treated culture [136]. At much lower concentrations (3×10^{-6} M), however, ergosta-5,7,22,24(28)-tetraenol was the major product, indicating that $\Delta^{24(28)}$ reductase is highly sensitive to the drug [137].

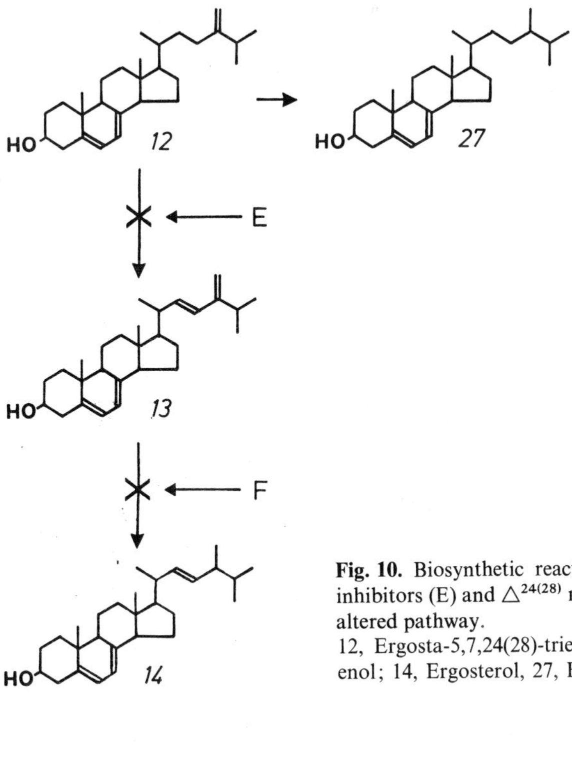

Fig. 10. Biosynthetic reactions blocked by Δ^{22} desaturation inhibitors (E) and $\Delta^{24(28)}$ reduction inhibitors (F) and resultant altered pathway.
12, Ergosta-5,7,24(28)-trienol; 13, Ergosta-5,7,22,24(28)-tetraenol; 14, Ergosterol, 27, Ergosta-5,7-dienol

Triarimol has been pointed out to interfere with the introduction of the 22(23) double bond in addition to its inhibition of 14α-demethylation, since ergosta-5,7-dienol accumulated in the treated culture of *A. fumigatus* [65] (Fig. 10). The Δ^{22} desaturation was reported to be mediated by a cytochrome P-450 containing monooxygenase system in *S. cerevisiae* [138]. The form of cytochrome P-450 involved in this reaction presumably differs from that responsible for 14α-demethylation, because Δ^{22} desaturation-deficient mutants of fungi are known to produce 14α-demethylated sterols [16,24]. Triarimol therefore possibly attacks both forms of the cytochrome P-450 in fungi. It is interesting to note in this connection that fenarimol is more toxic to a Δ^{22} desaturation-deficient mutant of *A. nidulans* than to the wild-type [24]. Genetic impairment for producing cytochrome P-450 involved in Δ^{22} desaturation may influence the molecular structure of cytochrome P-450 involved in 14α-demethylation, resulting in a higher affinity of fenarimol to the latter cytochrome P-450.

5 Miscellaneous Effects of Sterol-biosynthesis Inhibiting Fungicides

5.1 Effects on Membrane Permeability

Antimycotic drugs, clotrimazole and miconazole are fungicidal at high concentrations but fungistatic at low concentrations. The number of viable cells of *S. cerevisiae* was rapidly decreased by both compounds at a high concentration (1×10^{-4} M) [91,139]. Iwata et al. [140] found by employing *C. albicans* that clotrimazole at fungicidal concentrations (5×10^{-5} to 1.5×10^{-4} M) inhibited the incorporation of radioactive substrates into protein, nucleic acids and other cellular constituents, and it caused the leakage of ^{32}P-labelled substances and potassium from the cells. Miconazole had similar effects on the permeability of cell membranes in *C. albicans* at fungicidal concentrations (5×10^{-5} to 2×10^{-4} M) as demonstrated by the leakage of 260 nm-absorbing materials, and potassium and sodium ions [141]. These permeability changes were also found by utilizing lecithin liposomes. The imidazole antimycotics induced release of trapped glucose from the liposomes prepared from egg lecithin and dioleoyl lecithin [142]. But this effect was not prominent when dipalmitoyl lecithin liposomes were treated. It is likely that the permeability changes of liposomes caused by the imidazoles were more pronounced in the presence of unsaturated fatty acyl groups in the lecithin. Furthermore, the incorporation of unsaturated free fatty acids into egg lecithin liposomes increased the sensitivity to the imidazole drugs [143]. Liposomes prepared from phospholipids of an unusual colorless algae *Prototheca wickerhamii* likewise became more sensitive to miconazole and clotrimazole by the addition of free fatty acids of the same organism which contained unsaturated fatty acids at a high proportion [144]. The imidazole antimycotis, therefore, appear to interact with cell membranes, probably with unsaturated fatty acids. However, this is not always the case, since miconazole induces potassium release from a gram-positive bacterium *Staphylococcus aureus* which possesses only saturated fatty acids [145]. This suggests that there are other targets for imidazole membrane damage.

Divalent cations, Mg^{++} and Ca^{++} protected *C. albicans* from the fungitoxic effect of miconazole and diminished the leakage of 260 nm-absorbing materials from the cells [141,146]. This was supposed to be due to a competition between the ions and miconazole in its positively charged form for negatively charged binding sites [146]. Protection of *C. albicans* from the toxic effect of clotrimazole and miconazole was also attained by unsaturated fatty acids, and phospholipids and acyl glycerides having unsaturated fatty acid moieties [147,148]. The inhibition of leucine uptake of *C. albicans* by clotrimazole and miconazole was partially prevented by egg lecithin [85]. Similarly, unsaturated free fatty acids reduced growth inhibitory effects of miconazole against *S. aureus* [149]. The mechanism of annulment of the imidazole toxicity by unsaturated fatty acids is not yet fully elucidated, although it was reported that the physicochemical interaction of the imidazoles with unsaturated fatty acids might occur to form insoluble complexes in the ambient solutions [14]

Ketoconazole differs from the other imidazole antimycotics, since it is not fungicidal against *S. cerevisiae* [139] and does not cause release of K^+ from *C. albicans* [150]. A potent fungitoxic effect of the drug can be simply attributed to the inhibition of ergosterol biosynthesis [96,97]. In contrast, clotrimazole and miconazole possess at least two distinct antifungal actions. At lower concentrations, the imidazoles inhibit ergosterol biosynthesis and this ultimately results in retardation of fungal growth, whereas immediate disruption of fungal membranes occurred at higher concentrations [91]. The latter effect might be the major reason for the imidazoles being fungicidal and bactericidal in laboratory experiments. And this may explain why both clotrimazole and miconazole inhibited the growth of the sterol auxotrophic mutant and the C-14 demethylase-deficient mutant of *S. cerevisiae* [151]. Such membrane damages might occur because they probably act as cationic surface active agents like dodine, an agricultural fungicide. Tridemorph at a high concentration (1×10^{-4} M) caused leakage of ^{14}C-labelled substances from *B. cinerea* cells, as the chemical structure suggests that it acts as a cationic surface active agent. But this did not occur at a low concentration (1×10^{-5} M) of the fungicides, which was still very effective in inhibiting ergosterol synthesis [101]. Since higher concentrations of the fungicides are usually required for direct membrane damage, it is reasonable to consider that the inhibition of ergosterol biosynthesis is of primary importance in exerting fungitoxic effects for these fungicides. This was further supported by the observation that the C-14 demethylase-deficient mutant of U. maydis exhibits resistance to miconazole [152].

5.2 Morphological Changes

All of the 14α-demethylation inhibitors cause typical morphological changes in fungi. When *U. maydis* or *U. avenae* was cultured in the presence of the inhibitors, sporidial multiplication was severely prevented and the treated sporidia did not separate daughter cells, becoming multicellular and enlarged [50,69,73,78,80,85]. Morphology of filamentous fungi was also altered by these inhibitors. On cucumber leaves sprayed with buthiobate at concentrations ranging from 1 to 10^3 μg/ml, conidia of a powdery mildew fungus *Sphaerotheca fuliginea* germinate normally, but the germ tubes ceased to extend further and their tips became abnormally swollen [57]. Conidia of *M. fructigena* and *C. cucumerinum* germinate in the fungitoxic concentrations of buthiobate and triforine, respectively, but the germ tubes were distorted, swollen and excessively, branched [58,64]. Similarly, imazalil only weakly inhibited spore germination of *Penicillium italicum*, but the treated conidia were vacuolated and produced distorted germ tubes [153]. When *T. deformans*, which grows as an yeast form in the culture, was treated with propiconazole, the cells became swollen and vacuolated, and often formed clusters [154]. Clusters of interconnected yeast cells were also produced, when *C. albicans* was grown in the presence of miconazole [155] and bifonazole [156]. The latter compound has been regarded as a 14α-demethylation inhibitor because of its chemical structure (Fig. 4).

The effects of the imidazole antimycotics on the fungal ultrastructure markedly differed, depending on the treated concentrations. At fungicidal concentrations (1.5×10^{-4} M), clotrimazole caused rapid disruption of the nuclear envelope and other membranes in *C. albicans* [157]. The protoplasmic membranes often invaginated to form multivesicular structures and plasmolysis was ultimately observed [158]. When sphaeroplasts of *C. albicans* were directly treated with clotrimazole at 1×10^{-3} M, numerous circular openings were formed in the cytoplasmic membranes [159]. Econazole also caused severe membrane damage in *S. cerevisiae* at a fungicidal concentration (100 μg per ml) [160]. Degeneration of cytoplasmic organelles in *C. albicans* was noted to occur after exposure to a fungicidal concentration (1×10^{-4} M) of miconazole [161]. Scanning electron microscopy indicated that *C. albicans* cells exposed to 1×10^{-4} M miconazole were covered with numerous vesicles on smooth-walled cells which seemed to be derived from broken cells [155]. These morphological observations are therefore consistent with the results of biochemical studies which indicated the permeability changes of fungal membranes in the presence of fungicidal concentrations of the imidazoles.

At fungistatic concentrations, however, noticeable changes occurred at the periphery of the treated cells after 24 hrs-exposure to the imidazoles. *C. albicans* cells exposed to 1×10^{-7} M miconazole produced clusters of interconnected cells having a rough and wrinkled surface of the cell wall [155]. The cell wall appeared to be thickened and small vesicular inclusions were embedded in the wall [155, 161, 162]. Vesiculation of the plasmalemma and segregation of cytoplasmic material were sometimes recognized at the cell periphery [162]. Similar wall thickening and appearance of inclusions in the wall were noted when *Coccidioides immitis* was treated with miconazole and ketoconazole [163]. Observations of *C. albicans* and *Torulopsis glabrata* treated with bifonazole by freeze fracture electron microscopy gave more information on the morphological changes of plasmalemma [164]. In control cells of both yeasts, the numerous regularly shaped invaginations appeared in the convex protoplasmic fracture surface. But after 24-hrs exposure to bifonazole, the membrane invaginations became large and irregular and decreased in number. The plasmalemma and cell wall did not fit closely together, and small vesicles appeared between them.

The imidazole antimycotics therefore appeared to affect fungal membranes in at least two ways. At fungicidal concentrations, disruption of the membranes occurred immediately after the exposure to the drugs possibly by disorganizing the integrity of phospholipid bilayers. The fungistatic lower concentrations of the imidazoles do not show such rapid effects on fungal membranes but unusual membrane systems developed during a long exposure to the drugs, probably through the deficiency of ergosterol as an essential membrane component.

Ultrastructures of phytopathogenic fungi, *U. avenae*, *Botrytis allii* and *Uromyces vicia-fabae* were also tremendously altered by the treatment with the fungicides inhibiting 14α-demethylation. The cell wall of *U. avenae* became thickened in the presence of triadimefon, nuarimol and imazalil [165, 166]. Small vesicles were formed between the plasmalemma and cell wall, and the septa remained incomplete. *B. allii* and *U. vicia-fabae* treated with triadimefon showed principally similar morphological changes; the incomplete septa, the unevenly thickened cell wall and membranous inclusions in the thickened wall [167, 168].

It is therefore likely that all 14α-demethylation inhibitors at fungistatic concentrations exert similar or identical effects on fungal ultrastructures in a broad range of fungal species. Thickening of the cell wall, formation of small vesicles between plasmalemma and the cell wall and deposition of vesicular materials in the thickened wall were commonly observed in the treated cells. It is interesting to note however that naftifine [169] (Fig. 11), an inhibitor of squalene epoxidase [170], caused morphological changes in *Trichophyton metagrophytes* [171] and *Candida parapsilosis* [172], which resembled those observed in the fungi treated with 14α-demethylation inhibitors.

The mechanisms for abnormal deposition in the cell wall and development of unusual membrane systems in the treated fungi still remain obscure. However, it is reasonable to postulate that the deficiency of ergosterol and the accumulation of the sterol intermediates in the membranes accelerate development of unusual membrane systems and disorganize the regulation of cell wall construction in the membranes.

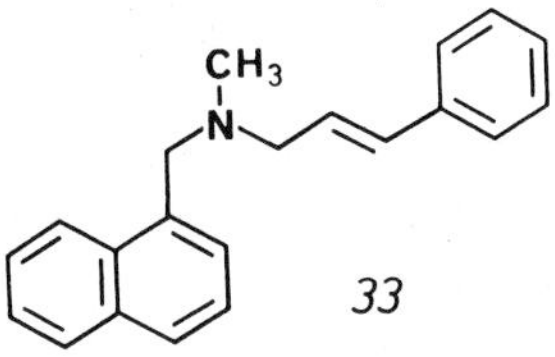

33

Fig. 11. Chemical structure of naftifine

6 Conclusion

Fungicides inhibiting 14α-demethylation are a big commercial success, because of their high performance at low dosages. It is surprising that a fairly broad range of compounds with diverse chemical structures inhibit 14α-demethylation in the conversion of lanosterol to ergosterol in fungi. Nevertheless, similarities in chemical structures are recognized among the compounds with the exception of triforine; (1) they are substituted aromatic heterocycles, (2) a nitrogen atom is located at the third position in the heterocycles, and (3) a benzene ring substituted with lipophilic group(s) is attached to the molecules. These inhibitors must attack cytochrome P-450 involved in 14α-demethylation, since at least one representative compound buthiobate was proven to interfere with the yeast cytochrome P-450 [59]. It is likely that the *meta*-nitrogen atom of the aromatic heterocycles binds to the protohaem iron of cytochrome P-450, inducing a Type II spectral change. The presence of the nitrogen atom at the third position is absolutely necessary for fungitoxicity [173] and to inhibit 14α-demethylation in the yeast cell-free system [174], whereas the lipophilic part of the inhibitors is structurally variable to a great extent. This may indicate that a coordinate bond must be formed in an exact arrangement between the nitrogen atom and the protohaem iron, but binding of the other part of the molecules occurs at various positions through the formation of hydrophobic bonds and in some cases hydrogen bonds.

Triforine largely differs from the other 14α-demethylation inhibitors in chemical structure. The fungicide did not inhibit 14α-demethylation in cell-free homogenates of *S. cerevisiae* [102], although it effectively prevented the conversion of 24-methylenedihydrolanosterol to ergosterol in whole cells of *N. crassa* and *A. fumigatus* [64,65]. A morphological effect of triforine against *S. fuliginea* was distinct from that of the other 14α-demethylation inhibitors [102]. These observations indicate a possibility that triforine inhibits 14α-demethylation by a different molecular mechanism.

In the presence of 14α-demethylation inhibitors, 24-methylenedihydrolanosterol, obtusifoliol and 14α-methyl-ergosta-8,24(28)-dienol commonly accumulated instead of ergosterol in fungal cells. All of the accumulated sterols possess the 14α-methyl group which has been thought to be detrimental to stabilizing membranes [175]. In strictly anaerobic culture conditions, lanosterol cannot support the growth of *S.*

cerevisiae in the absence of ergosterol [176]. These observations are consistent with the idea that fungitoxicity of 14α-demethylation inhibitors can be ascribed to their ability to inhibit ergosterol biosynthesis. However, it was not possible to annul the fungitoxicity of 14α-demethylation inhibitors by adding ergosterol to the culture medium. The reason for this not fully explained. It is, however, suspected that 24-methylenedihydrolanosterol accumulating in a large amount occupies the positions for ergosterol in membrane structures, preventing the entrance of externally applied sterols. Localization of sterol intermediates in membranes of the treated cells can be assumed by the observation that the fungi cultured in the presence of 14α-demethylation inhibitors become resistant to polyene antibiotics [177], since polyenes are most toxic when ergosterol is present. Moreover, in the strictly anaerobic and ergosterol-supplemented culture of *S. cerevisiae*, the fungitoxicity of 14α-demethylation inhibitors was significantly diminished [91], probably because squalene is not oxidized and any cyclization intermediates are not present in the culture, and hence the entrance of ergosterol into membrane structures may not be disturbed.

$\triangle^8 \rightarrow \triangle^7$ Isomerization inhibitors are also an important segment of ergosterol biosynthesis inhibitors. Tridemorph and fenpropemorph are substituted morpholines and trifulperidol is a substituted piperidine. It can be pointed out that all of this type of inhibitors are secondary or tertiary amines substituted with fairly lipophilic groups (Fig. 5). The $\triangle^8 \rightarrow \triangle^7$ isomerase localizes in microsomes of *S. cerevisiae* [178] and lipophilic alkyl amines may be easily incorporated into such membranous parts of the cells to disturb the reaction.

Most of the other sterol biosynthesis inhibitors described in this chapter possess analogous structures to normal sterols (Fig. 7). They seem to act as competitive inhibitors in the reactions involved.

The accumulated sterols in the fungal cells treated with these inhibitors are incompletely demethylated and thought to replace ergosterol. The reason for the fungitoxicity of the inhibitors is not sufficiently understood, but such unusual intermediates probably cause small changes in the precise molecular architecture of membranes, inducing changes of membrane properties, and ultimately resulting in inevitable defects in fungal growth.

In the conversion of lanosterol to ergosterol, inhibitors of C-24 transmethylation, 14α-demethylation, $\triangle^8 \rightarrow \triangle^7$ isomerization, $\triangle^{22}$ desaturation, $\triangle^{24(28)}$ reduction and $\triangle^{14}$ reduction have been found as described in this chapter. However, we do not know of antifungal compounds which specifically inhibit $\triangle^5$ desaturation or demethylation at C-4. It would not be surprising if such inhibitors were to appear in the future. Ergosterol biosynthesis is still a very attractive target for future fungicides with diverse chemical structures.

7 References

1. Bird, C. W., Lynch, J. M., Pirt, F. J., Reid, W. W., Brooks, C. J. W. and Middleditch, B. S.: Nature *230*, 473 (1971)
2. Smith, P. F.: J. Lipid Res. *5*, 121 (1964)
3. Ourisson, G., Albrecht, P. and Rohmer, M.: Pure Appl. Chem. *51*, 709 (1979)

4. Andreasen, A. A. and Stier, T. J. B.: J. Cell. Comp. Physiol. *41*, 23 (1953)
5. Morpurgo, G., Serlupi-Crescenzi, G., Tecce, G., Valente, F. and Venettacci, D.: Nature *201*, 897 (1964)
6. Wallace, P. G., Huang, M. and Linnane, A. W.: J. Cell Biol. *37*, 207 (1968)
7. Hamilton-Miller, J. M. T.: Adv. Appl. Microbiol. *17*, 109 (1974)
8. Bean, G. A.: Adv. Lipid Res. *11*, 193 (1973)
9. Weete, J. D.: Phytochemistry *12*, 1843 (1973)
10. Weete, J. D. and Laseter, J. L.: Lipids *9*, 575 (1974)
11. Hung-Kuang, L., Langenbach, R. J. and Knoche, H. W.: Phytochemistry *11*, 2319 (1972)
12. McCorkindale, N. J., Hutchinson, S. A., Pursey, B. A., Scott, W. T. and Wheeler, R.: Phytochemistry *8*, 861 (1969)
13. Hendrix, J. W.: Annu. Rev. Phytopathol. *8*, 111 (1970)
14. Warner, S. A., Sovocool, G. W. and Domnas, A. J.: Phytochemistry *21*, 2135 (1982)
15. Parks, L. W., Bond, F. T., Thompson, E. D. and Starr, P. R.: J. Lipid Res. *13*, 311 (1972)
16. Barton, D. H. R., Corrie, J. E. T., Widdowson, D. A., Bard, M. and Woods, R. A.: J. C. S. Perkin I *1974*, 1326
17. Trocha, P. J., Jasne, S. J. and Sprinson, D. B.: Biochem. Biophys. Res. Commun. *59*, 666 (1974)
18. Bard, M., Woods, R. A., Barton, D. H. R., Corrie, J. E. T. and Widdowson, D. A.: Lipids *12*, 645 (1977)
19. Subden, R. E., Safe, L., Morris, D. C., Brown, R. G. and Safe, S.: Can. J. Microbiol. *23*, 751 (1977)
20. Pierce, A. M., Pierce, H. D., Jr., Unrau, A. M. and Oehlschlager, A. C.: Can. J. Biochem. *56*, 135 (1978)
21. Kim, S. J., Kwon-Chung, K. J., Milne, G. W. A. and Prescott, B.: Antimicrob. Ag. Chemother. *6*, 405 (1974)
22. Grindle, M. and Farrow, R.: Molec. Gen. Genet. *165*, 305 (1978)
23. Walsh, R. C. and Sisler, H. D.: Pestic. Biochem. Physiol. *18*, 122 (1982)
24. Ziogas, B. N., Sisler, H. D. and Lusby, W. R.: Pestic. Biochem. Physiol. *20*, 320 (1983)
25. Bloch, K.: Science *150*, 19 (1965)
26. Gaylor, J. L.: Enzymes of sterol biosynthesis, in: Biochemistry of Lipids Vol. 4 (ed.) Goodwin, T. W., p. 1, University Park Press, Baltimore 1974
27. Nes, W. R.: Adv. Lipid Res. *15*, 233 (1977)
28. Fryberg, M., Oehlschlager, A. C. and Unrau, A. M.: J. Am. Chem. Soc. *95*, 5747 (1973)
29. Barton, D. H. R., Corrie, J. E. T., Marshall, P. J. and Widdowson, D. A.: Bioorg. Chem. *2*, 363 (1973)
30. Goulston, G., Mercer, E. I. and Goad, L. J.: Phytochemistry *14*, 457 (1975)
31. Epstein, W. W. and Lear, G. V.: J. Org. Chem. *31*, 3434 (1966)
32. Kato, T., Tanaka, S., Ueda, M. and Kawase, Y.: Agr. Biol. Chem. *39*, 169 (1975)
33. Katsuki, H. and Bloch, K.: J. Biol. Chem. *242*, 222 (1967)
34. Parks, L. W., Anding, C. and Ourisson, G.: Eur. J. Biochem. *43*, 451 (1974)
35. Fryberg, M., Oehlschlager, A. C. and Unrau, A. M.: Arch. Biochem. Biophys. *173*, 171 (1975)
36. Pierce, A. M., Mueller, R. B., Unrau, A. M. and Oehlschlager, A. C.: Can. J. Biochem. *56*, 794 (1978)
37. Parks, L. W.: J. Am. Chem. Soc. *80*, 2042 (1958)
38. Alexander, K., Akhtar, M., Boar, R. B., McGhie, J. F. and Barton, D. H. R.: J. C. S. Chem. Comm. *1972*, 383
39. Alexander, K. T. W., Mitropoulos, K. A. and Gibbons, G. F.: Biochem. Biophys. Res. Commun. *60*, 460 (1974)
40. Aoyama, Y. and Yoshida, Y.: Biochem. Biophys. Res. Commun. *82*, 33 (1978)
41. Aoyama, Y. and Yoshida, Y.: Biochem. Biophys. Res. Commun. *85*, 28 (1978)
42. Ohba, M., Sato, R., Yoshida, Y., Nishino, T. and Katsuki, H.: Biochem. Biophys. Res. Commun. *85*, 21 (1978)
43. Yoshida, Y. and Aoyama, Y.: J. Biol. Chem. *259*, 1655 (1984)
44. Aoyama, Y., Yoshida, Y. and Sato, R.: J. Biol. Chem. *259*, 1661 (1984)
45. Parks, L. W.: CRC Crit. Rev. Microbiol. *6*, 301 (1978)
46. Nishino, T., Hata, S., Osumi, T. and Katsuki, H.: J. Biochem. *88*, 247 (1980)
47. Brown, I. F., Jr., Hall, H. R. and Miller, J. R.: Phytopathology *60*, 1013 (1970)

48. Brown, I. F., Jr. and Hall, H. R.: Phytopathology *61*, 886 (1971)
49. Ragsdale, N. N. and Sisler, H. D.: Biochem. Biophys. Res. Commun. *46*, 2048 (1972)
50. Ragsdale, N. N. and Sisler, H. D.: Pestic. Biochem. Physiol. *3*, 20 (1973)
51. Ragsdale, N. N.: Biochim. Biophys. Acta *380*, 81 (1975)
52. Kato, T. and Kawase, Y.: Agr. Biol. Chem. *40*, 2379 (1976)
53. Mitropoulos, K. A. and Gibbons, G. F.: Biochem. Biophys. Res. Commun. *71*, 892 (1976)
54. Frasinel, C., Patterson, G. W. and Dutky, S. R.: Phytochemistry *17*, 1567 (1978)
55. Buchenauer, H.: Z. PflKrankh. PflSchutz *84*, 286 (1977)
56. Schmitt, P. and Benveniste, P.: Phytochemistry *18*, 1659 (1979)
57. Kato, T., Tanaka, S., Yamamoto, S., Kawase, Y. and Ueda, M.: Ann. Phytopathol. Soc. Japan *41*, 1 (1975)
58. Kato, T., Tanaka, S., Ueda, M. and Kawase, Y.: Agr. Biol. Chem. *38*, 2377 (1974)
59. Aoyama, Y., Yoshida, Y., Hata, S., Nishino, T. and Katsuki, H.: Biochem. Biophys. Res. Commun. *115*, 642 (1983)
60. Ebenebe, Ch., Fehrmann, H. and Grossmann, F.: Pl. Dis. Reptr *55*, 691 (1971)
61. Fuchs, A., Doma, S. and Vörös, J.: Neth. J. Pl. Path. *77*, 42 (1971)
62. Szkolnik, M.: Pl. Dis. Reptr *58*, 587 (1974)
63. Drandarevski, C. A. and Schicke, P.: Z. PflKrankh. PflSchutz *83*, 385 (1976)
64. Sherald, J. L., Ragsdale, N. N. and Sisler, H. D.: Pestic. Sci. *4*, 719 (1973)
65. Sherald, J. L. and Sisler, H. D.: Pestic. Biochem. Physiol. *5*, 477 (1975)
66. Brandes, W., Van Den Boom, T. and Kaspers, H.: Med. Fac. Landbouww. Rijksuniv. Gent *41*, 623 (1976)
67. Buchenauer, H.: Z. PflKrankh. PflSchutz *83*, 363 (1976)
68. Leroux, P. and Gredt, M.: Phytopath. Z.: *86*, 276 (1976)
69. Buchenauer, H.: Pestic. Biochem. Physiol. *7*, 309 (1977)
70. Gadher, P., Mercer, E. I., Baldwin, B. C. and Wiggins, T. E.: Pestic. Biochem. Physiol. *19*, 1 (1983)
71. Berg, D., Draber, W., von Hugo, H., Hummel, W. and Mayer, D.: Z. Naturforsch. *36c*, 798 (1981)
72. Jeffrey, R. A., Rowley, N. K. and Smailes, A.: Proc. Brit. Insectic. Fungic. Conf. *1975*, 429
73. Buchenauer, H.: Pestic. Biochem. Physiol. *8*, 15 (1978)
74. Urech, P. A., Schwinn, F. J., Speich, J. and Staub, Th.: Proc. Brit. Crop Prot. Conf. Pests Dis. *1979*, 508
75. Henry, M. J. and Sisler, H. D.: Pestic. Sci. *12*, 98 (1981)
76. Ebert, E., Gaudin, J., Muecke, W., Ramsteiner, K., Vogel, C. and Fuhrer, H.: Z. Naturforsch. *38c*, 28 (1983)
77. Weete, J. D., Sancholle, M. S. and Montant, C.: Biochim. Biophys. Acta *752*, 19 (1983)
78. Takano, H., Oguri, Y. and Kato, T.: J. Pestic. Sci. *8*, 575 (1983)
79. Bent, K. J. and Skidmore, A. M.: Proc. Brit. Crop Prot. Conf. Pests Dis. *1979*, 477
80. Kraus, P.: PflSchutz-Nachr. (Bayer) *32*, 17 (1979)
81. Gasztonyi, M. and Josepovits, G.: Pestic. Sci. *10*, 57 (1979)
82. Buchenauer, H. and Grossmann, F.: Z. PflKrankh. PflSchutz *89*, 309 (1982)
83. Wainwright, A., Rollett, A. C. and Morris, D. B.: Proc. Brit. Crop Prot. Conf. Pests Dis. *1979*, 565
83-1. Skidmore, A. M., French, P. N. and Rathmell, W. G.: Proc. 10th Int. Cong. Pl. Prot. *1983*, 368
83-2. Eberle, J., Ruess, W. and Urech, P. A.: Proc. 10th Int. Cong. Pl. Prot. *1983*, 376
83-3. Espinel-Ingroff, A., Shadomy, S. and Gebhart, R. J.: Antimicrob. Ag. Chemother. *26*, 5 (1984)
83-4. Yamaguchi, H. and Hiratani, T.: J. Antimicrob. Chem. *11*, 135 (1983)
84. Laville, E.: Fruits *28*, 545 (1973)
85. Buchenauer, G.: Z. PflKrankh. PflSchutz *84*, 440 (1977)
86. Leroux, P. and Gredt, M.: Ann. Phytopath. *10*, 45 (1978)
87. Leroux, P. and Gredt, M.: C. R. Acad. Sc. Paris, t. *286*, 427 (1978)
88. Siegel, M. R. and Ragsdale, N. N.: Pestic. Biochem. Physiol. *9*, 48 (1978)
89. Harris, R. G., Weighton, D. M., Blanquat, A. de St. and Rose, I. D. G.: Proc. Brit. Crop Prot. Conf. Pests Dis. *1979*, 53
90. Sawyer, P. R., Brogden, R. N., Pinder, R. M., Speight, T. M. and Avery, G. S.: Drugs *9*, 424 (1975)
91. Sud, I. J. and Feingold, D. S.: J. Invest. Dermatol. *76*, 438 (1981)

92. Sawyer, P. R., Brogden, R. N., Pinder, R. M., Speight, T. M. and Avery, G. S.: Drugs *9*, 406 (1975)
93. Van Den Bossche, H., Willemsens, G., Cools, W., Lauwers, W. F. J. and Le Jeune, L.: Chem. Biol. Interact. *21*, 59 (1978)
94. Henry, M. J. and Sisler, H. D.: Antimicrob. Ag. Chemother. *15*, 603 (1979)
95. Thienpont, D., Van Cutsem, J., Van Gerven, F., Heeres, J. and Janssen, P. A. J.: Experientia *35*, 606 (1979)
96. Van Den Bossche, H., Willemsens, G., Cools, W., Cornelissen, F., Lauwers, W. F. and Van Cutsem, J. M.: Antimicrob. Ag. Chemother. *17*, 922 (1980)
97. Marriott, M. S.: J. Gen. Microbiol. *117*, 253 (1980)
98. Jevons, S., Gymer, G. E., Brammer, K. W., Cox, D. A. and Leeming, M. R. G.: Antimicrob. Ag. Chemother. *15*, 597 (1979)
99. Thienpont, D., Van Cutsem, J., Van Nueten, J. M., Niemegeers, C. J. E. and Marsboom, R.: Arzneim.-Forsch. *25*, 224 (1975)
100. Pye, G. W. and Marriott, M. S.: Sabouraudia *20*, 325 (1982)
100-1. Barug, D. and Kerkenaar, A.: Pestic. Sci. *15*, 78 (1984)
100-2. Polak, A.: Arzneim.-Forsch. *32*, 17 (1982)
100-3. Tajana, A., Sibilia, C., Cappelletti, R., Cova, A. and Nardi., D.: Arzneim.-Forsch. *31*, 2127 (1981)
100-4. Iwata, K. and Yamamoto, Y.: Jpn. J. Med. Mycol. *23*, 314 (1982)
101. Kato, T., Shoami, M. and Kawase, Y.: J. Pestic. Sci. *5*, 69 (1980)
102. Kato, T.: Biosynthetic processes of ergosterol as the target of fungicides, in: Pesticide Chemistry, Human Welfare and the Environment Vol. 3 Mode of Action, Metabolism and Toxicology (eds.) Matsunaka, S., Hutson, D. H. and Murphy, S. D., p. 33, Pergamon Press, Oxford 1983
103. Bu'Lock, J. D. and Osagie, A. U.: Phytochemistry *15*, 1249 (1976)
104. Berg, L. R., Patterson, G. W. and Lusby, W. R.: Lipids *18*, 448 (1983)
105. Schmitt, P., Benveniste, P. and Leroux, P.: Phytochemistry *20*, 2153 (1981)
106. Bladocha, M. and Benveniste, P.: Plant Physiol. *71*, 756 (1983)
107. Kerkenaar, A., Uchiyama, M. and Versluis, G. G.: Pestic. Biochem. Physiol. *16*, 97 (1981)
108. Simpson, G. M. and Cooper, T. B.: Curr. Ther. Res. *8*, 249 (1966)
109. Sobus, M. T., Holmlund, C. E. and Whittaker, N. F.: J. Bacteriol. *130*, 1310 (1977)
110. Blohm, T. R. and MacKenzie, R. D.: Arch. Biochem. Biophys. *85*, 245 (1959)
111. Steinberg, D. and Avigan, J.: J. Biol. Chem. *235*, 3127 (1960)
112. Campagnoni, C., Holmlund, C. E. and Whittaker, N.: Arch. Biochem. Biophys. *184*, 555 (1977)
113. Dvornik, D., Kraml, M., Dubuc, J., Givner, M. and Gaudry, R.: J. Am. Chem. Soc. *85*, 3309 (1963)
114. Pereira, R., Holmlund, C. E. and Whittaker, N.: Lipids *18*, 545 (1983)
115. Taketani, S., Nagai, J. and Katsuki, H.: Biochim. Biophys. Acta *528*, 416 (1978)
116. Schmitt, P. and Benveniste, P.: Phytochemistry *18*, 445 (1979)
117. Bohnen, K., Siegle, H. and Löcher, F.: Proc. Brit. Crop Prot. Conf. Pests Dis. *1979*, 541
118. Boeck, L. D., Hoehn, M. M., Westhead, J. E., Wolter, R. K. and Thomas, D. N.: J. Antibiotics *28*, 95 (1975)
119. Hays, P. R., Parks, L. W., Pierce, H. D., Jr., and Oehlschlager, A. C.: Lipids *12*, 666 (1977)
120. Hays, P. R., Neal, W. D. and Parks, L. W.: Antimicrob. Ag. Chemother. *12*, 185 (1977)
121. Bottema, C. K. and Parks, L. W.: Biochim. Biophys. Acta *531*, 301 (1978)
122. Woloshuk, C. P., Sisler, H. D. and Dutky, S. R.: Antimicrob. Ag. Chemother. *16*, 81 (1979)
123. Schmitt, P., Scheid, F. and Benveniste, P.: Phytochemistry *19*, 525 (1980)
124. Dickson, L. G., Patterson, G. W., Cohen, C. F. and Dutky, S. R.: Phytochemistry *11*, 3473 (1972)
125. Dickson, L. G. and Patterson, G. W.: Lipids *7*, 635 (1972)
126. Gibbons, G. F. and Mitropoulos, K. A.: Biochim. Biophys. Acta *380*, 270 (1975)
127. Lutsky, B. N., Hsiung, H. M. and Schroepfer, G. J., Jr.: Lipids *10*, 9 (1975)
128. Dvornik, D., Kraml, M. and Bagli, J. F.: Biochemistry *5*, 1060 (1966)
129. Avruch, L., Fischer, S., Pierce, H., Jr. and Oehlschlager, A. C.: Can. J. Biochem. *54*, 657 (1976)
130. Kabara, J. J., Holzschu, D. L. and Catsoulacos, D. P.: Lipids *11*, 755 (1976)
131. Pierce, H. D., Jr., Pierce, A. M., Srinivasan, R., Unrau, A. M. and Oehlschlager, A. C.: Biochim. Biophys. Acta *529*, 429 (1978)
132. Pierce, A. M., Unrau, A. M. and Oehlschlager, A. C.: Can. J. Biochem. *57*, 201 (1979)
133. Bailey, R. B., Hays, P. R. and Parks, L. W.: J. Bacteriol. *128*, 730 (1976)

134. Sipe, J. D. and Holmlund, C. E.: Biochim. Biophys. Acta *280*, 145 (1972)
135. Fung, B. and Holmlund, C. E.: Biochem. Pharmacol. *25*, 1249 (1976)
136. Field, R. B. and Holmlund, C. E.: Arch. Biochem. Biophys. *180*, 465 (1977)
137. Field, R. B., Holmlund, C. E. and Whittaker, N. F.: Lipids *14*, 741 (1979)
138. Hata, S., Nishino, T., Komori, M. and Katsuki, H.: Biochem. Biophys. Res. Commun. *103*, 272 (1981)
139. Sud, I. J. and Feingold, D. S.: Antimicrob. Ag. Chemother. *20*, 71 (1981)
140. Iwata, K., Yamaguchi, H. and Hiratani, T.: Sabouraudia *11*, 158 (1973)
141. Swamy, K. H. S., Sirsi, M. and Rao, G. R.: Antimicrob. Ag. Chemother. *5*, 420 (1974)
142. Yamaguchi, H. and Iwata, K.: Antimicrob. Ag. Chemother. *15*, 706 (1979)
143. Sud, I. J., Chou, D.-L. and Feingold, D. S.: Antimicrob. Ag. Chemother. *16*, 660 (1979)
144. Sud, I. J. and Feingold, D. S.: Antimicrob. Ag. Chemother. *16*, 486 (1979)
145. Sud, I. J. and Feingold, D. S.: Antimicrob. Ag. Chemother. *22*, 470 (1982)
146. Cope, J. E.: J. Gen. Microbiol. *119*, 245 (1980)
147. Yamaguchi, H.: Antimicrob. Ag. Chemother. *12*, 16 (1977)
148. Yamaguchi, H.: Antimicrob. Ag. Chemother. *13*, 423 (1978)
149. Van Den Bossche, H., Ruysschaert, J. M., Defrise-Quertain, F. Willemsens, G., Cornelissen, F., Marichal, P., Cools, W. and Van Cutsem, J.: Biochem. Pharmacol. *31*, 2609 (1982)
150. Beggs, W. H.: J. Antimicrob. Chemother. *11*, 381 (1983)
151. Taylor, F. R., Rodriguez, R. J. and Parks, L. W.: Antimicrob. Ag. Chemother. *23*, 515 (1983)
152. Sisler, H. D., Walsh, R. C. and Ziogas, B. N.: Ergosterol Biosynthesis: a target of fungitoxic action, in: Pesticide Chemistry, Human Welfare and the Environment Vol. 3 Mode of Action, Metabolism and Toxicology (eds.) Matsunaka, S., Hutson, D. H. and Murphy, S. D., p. 129, Pergamon Press Oxford 1983
153. Siegel, M. R., Kerkenaar, A. and Kaars Sijpesteijn, A.: Neth. J. Pl. Path. *83* (Suppl. 1), 121 (1977)
154. Sancholle, M., Weete, J. D. and Montant, C.: Pestic. Biochem. Physiol. *21*, 31 (1984)
155. De Nollin, S. and Borgers, M.: Antimicrob. Ag. Chemother. *7*, 704 (1975)
156. Barug, D., Samson, R. A. and Kerkenaar, A.: Arzneim.-Forsch. *33*, 528 (1983)
157. Yamaguchi, H., Kanda, Y., Iwata, K. and Osumi, M.: J. Electron Microsc. *22*, 167 (1973)
158. Iwata, K., Kanda, Y., Yamaguchi, H. and Osumi, M.: Sabouraudia *11*, 205 (1973)
159. Bastide, M., Jouvert, S. and Bastide, J.-M.: Can. J. Microbiol. *28*, 1119 (1982)
160. Yamaguchi, H., Iwata, K., Nagano, M. and Osumi, M.: J. Electron Microsc. *30*, 305 (1981)
161. Borgers, M.: Rev. Infect. Dis. *2*, 520 (1980)
162. De Nollin, S. and Borgers, M.: Sabouraudia *12*, 341 (1974)
163. Borgers, M., Levine, H. B. and Cobb, J. M.: Sabouraudia *19*, 27 (1981)
164. Barug, D. and de Groot, C.: Arzneim.-Forsch. *33*, 538 (1983)
165. Hippe, S., Buchenauer, H. and Grossmann, F.: Z. PflKrankh. PflSchutz *87*, 423 (1980)
166. Hippe, S. and Grossmann, F.: Pestic. Sci. *13*, 447 (1982)
167. Richmond, D. V.: Pestic. Biochem. Physiol. *21*, 74 (1984)
168. Pring, R. J.: Pestic. Biochem. Physiol. *21*, 127 (1984)
169. Georgopoulos, A., Petranyi, G., Mieth, H. and Drews, J.: Antimicrob. Ag. Chemother. *19*, 386 (1981)
170. Paltauf, F., Daum, G., Zuder, G., Högenauer, G., Schulz, G. and Seidl, G.: Biochim. Biophys. Acta *712*, 268 (1982)
171. Meingassner, J. G., Sleytr, U. and Petranyi, G.: J. Invest. Dermatol. *77*, 444 (1981)
172. Meingassner, J. G. and Sleytr, U. B.: Sabouraudia *20*, 199 (1982)
173. Tanaka, S., Kato, T., Yamamoto, S. and Yoshioka, H.: Agric. Biol. Chem. *41*, 1953 (1977)
174. Mercer, E. I.: Biochem. Soc. Transact. *11*, 663 (1983)
175. Bloch, K.: Curr. Topics Cell. Regul. *18*, 289 (1981)
176. Nes, W. R., Sekula, B. C., Nes, W. D. and Adler, J. H.: J. Biol. Chem. *253*, 6218 (1978)
177. Sud. and Feingold, D. S.: Antimicrob. Ag. Chemother. *23*, 185 (1983)
178. Yabusaki, Y., Nishino, T., Ariga, N. and Katsuki, H.: J. Biochem. *85*, 1531 (1979)

Chemistry of Sterol-biosynthesis Inhibiting Fungicides

Wolfgang Krämer
PF F Wirkstoff Forschung, Bayer AG, 5600 Wuppertal, FRG

Fungicides that interfere in sterol biosynthesis form one of the most important classes of fungicides and antimycotics today. With respect to their chemical structures they are very heterogenous. Nevertheless most of them can be grouped into three major classes: Pyrimidines and Pyridines, Azoles and Morpholines. The aim of the following article is to give an overview of the chemistry of these classes, to discuss structure activity relations within one chemical group of compounds, to show the importance of steric parameters for activity within these types of fungicides and to survey market products, and those at the development stage, with the corresponding syntheses, in crop protection and medicine.

The rapid worldwide progress in research on fungicides and antimycotics interfering in sterol biosynthesis makes it very difficult to give a complete description of this research area. Nevertheless this article should demonstrate not only the heterogenicity of the chemistry of such compounds but also their common structural requirements.

1 Introduction

The fungicides that interfere in sterol biosynthesis [1] are very heterogeneous with respect to their chemical structures. Nevertheless most can be grouped into three major classes:
Pyrimidines and Pyridines with hydroxy functions [2],
Azoles [3], and
Morpholines [4].

There are other individual agrochemical products from the piperazine-chloral-aminal [5,6] and N-3-pyridyl-dithiocarbonimidate classes [7], als well as the antimycotic naphthalene-methanamines [8], and the polyene antibiotics, azasterols, and sinefungine [9]. The azole group (derivatives of lH-imidazole and lH-triazole) has been the subject of the most intensive synthetic work and has also been widely surveyed [10].

The following article gives an overview of the chemistry of these compound classes, structure-activity relationships and stereochemical aspects with respect to biological activity.

2 Piperazines

Aminals of chloral [11] with the general formula

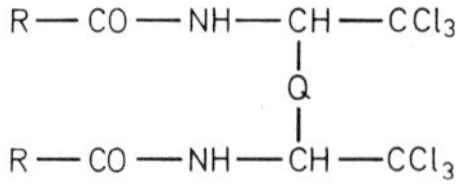

in which Q represents a piperazine or diamine residue, were first described as systemic fungicides in 1969 [12].

N,N¹-bis-(1-formamido-2,2,2-trichloroethyl)-piperazine of the structure

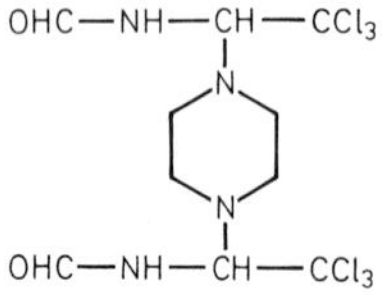

Triforine
Saprol®, Funginex®
Boehringer und Sohn
Celamerck GmbH
$LD_{50_{\text{rat, oral, acute}}}$: >16000 mg/kg

is prepared by reaction of (1,2,2,2-tetrachloroethyl)-formamide with piperazine in the presence of a base such as triethylamine or soda in acetone or water.

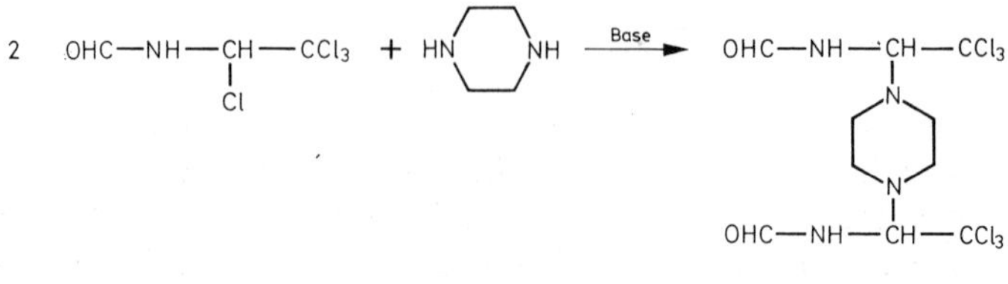

The compound has a melting point of 175° and is only slightly soluble in water (30 ppm) and hydrocarbon solvents (200 ppm).

3 Pyridines

Two subgroups may be distinguished within the pyridine class. Firstly the N-3-pyridyldithiocarbonimidates [13] which include the fungicide Denmert® (S-1358, Buthiobate), and secondly the 3-substituted pyridines of the Parnon® type (EL-241, Parimol), which are pyridine analogues of hydroxypyrimidines such as Rubigan®.

S-n-butyl-S¹-p.-tert.-butylbenzyl-N-3-pyridyldithiocarbonimidate [14,15]

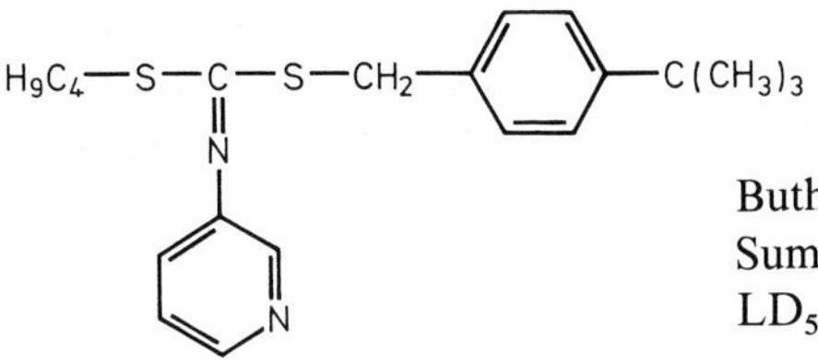

Buthiobate, Denmert®
Sumitomo Chem. Co. Ltd.
$LD_{50_{\text{rat, oral, acute}}}$: 2700–4900 mg/kg

is prepared by reaction of 3-aminopyridine with carbon disulphide then subsequent alkylation with N-butyl iodide and p-tert.-butylbenzyl bromide according to the following reaction sequence:

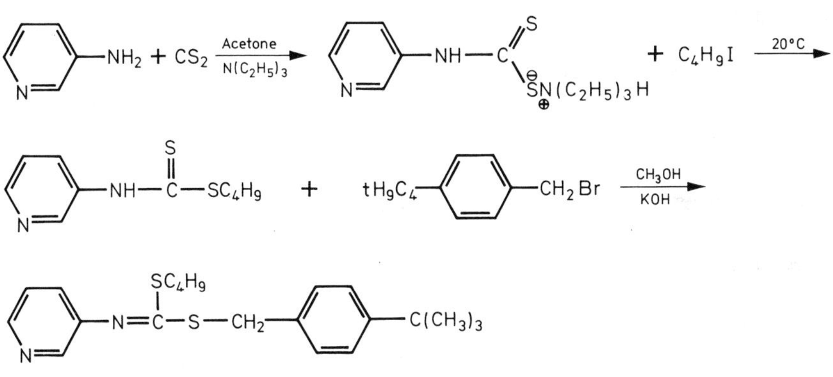

The end-product has a melting point of 31–33° and is practically insoluble in water but freely soluble in most organic solvents. The biological activity of the S-ethyl analogue is comparable to that of buthiobate. The corresponding derivatives of 2- and 4-aminopyridine are biologically inactive.

Structure-activity investigations [16] have shown that the fungicidal activity correlates with steric and lipophilic (R_m) parameters with respect to both the alkyl residue and the substitutents in the aromatic ring. Replacement of the S-butyl group by NH-ethyl also leads to biologically active compounds. Of the substituents studied in the pyridyl residue, only 5-methyl substitution provided biologically active derivatives. Replacement of the pyridyl ring by other heterocycles led to inactive products with the exception of pyrazine. Monothiocarbonamidates [17] in which either the thiobutyl function has been replaced by O-n-butyl, or the thiobenzyl function by O-benzyl, or thioformamidates in which the S-butyl or S-benzyl function has been replaced by CH_2-iC_3H_7 or CH_2-CH_2-Ph-C$(CH_3)_3$ respectively are also biologically active. To summarise, the general structure essential for fungicidal activity in compounds of the buthiobate type may be formulated as:

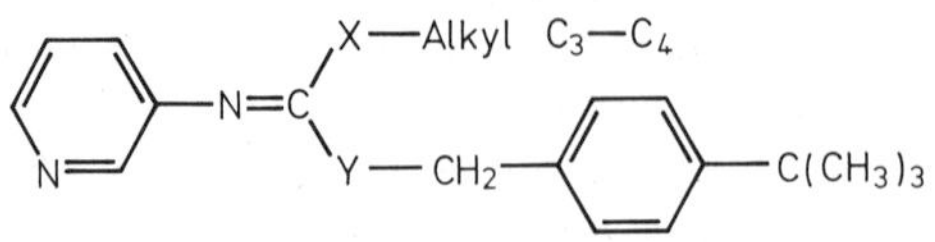

X = S ; O ; CH_2 ; NH

Y = S ; O ; CH_2

Further Denmert® analogues are described in the following patent applications:
NL 72 05 861 (1971)
J 50 046 832 (1974)
J 51 118 832 (1975)
J 50 094 133 (1974)

Various fungicides, some systemic, have been reported from the pyridine-alcohol [18] class researched extensively at Eli Lilly.

Bis-(p-chloro-phenyl)-3-pyridylmethanol [19]

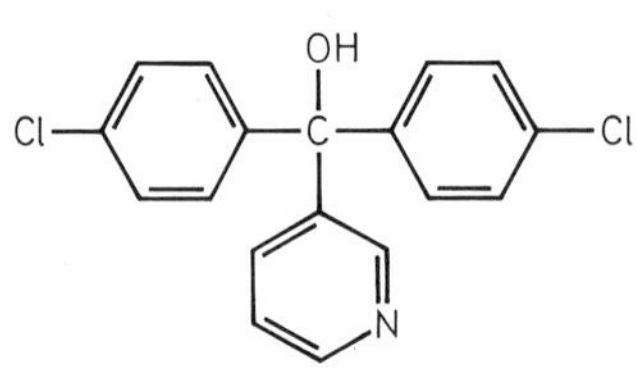

EL 241
Parinol, Parnon®
Eli Lilly Co.
m.p. 165–175 °C (as HCl-salt)

α-Cyclohexyl-α-phenyl-3-pyridinemethanol [20]

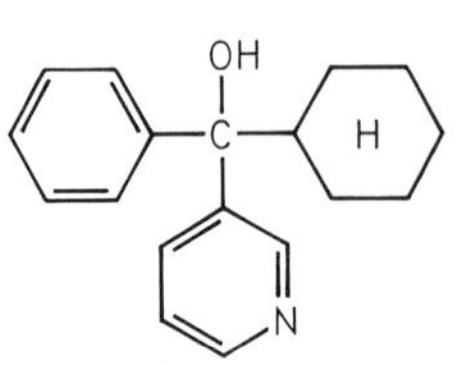

EL 331

These compounds are synthesized from the corresponding ketones by an organometallic reaction.

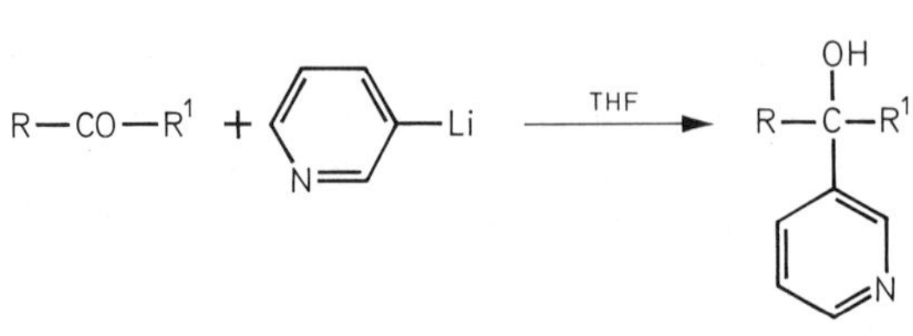

Alternatively, diaryl-pyridinemethanols can also be made in good yield by oxidation of the corresponding diaryl-methylpyridines [21]:

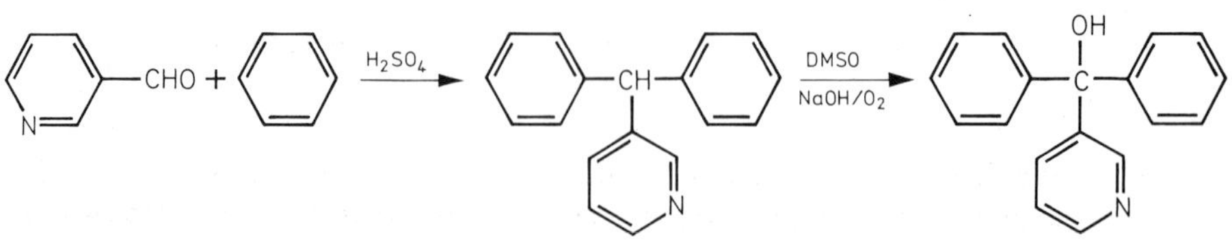

Substitution of the pyridine ring in the 3-position is the essential structural element for the fungicidal activity of the pyridinemethanols. The phenyl rings on the central carbon atom can be replaced by cycloalkyl residues, particulary cyclohexyl, with enhancement of activity. No significant increase in activity is observed upon wide-ranging variation of substituents in the phenyl rings. Replacement of the hydrogen or hydroxy group on the α-C-atom by chloro, amino or methoxy groups leads to loss of activity as does further substitution in the pyridine rings.

Further pyridine derivatives have been described as fungicides by various companies, e.g. *2-benzyl-2-(3-pyridyl)-1,3-dioxolanes* of structure *I* [22a] and *1-α-halo-benzyl-pyridine-ethanols* [22b] such as structure II:

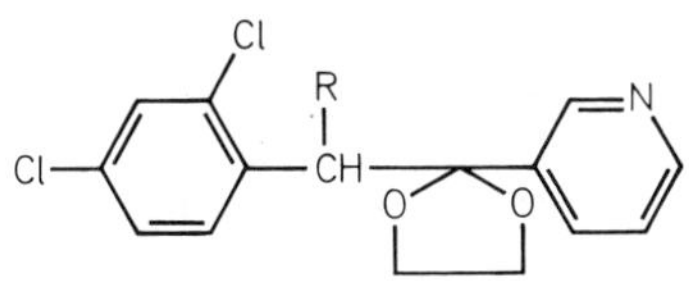

I

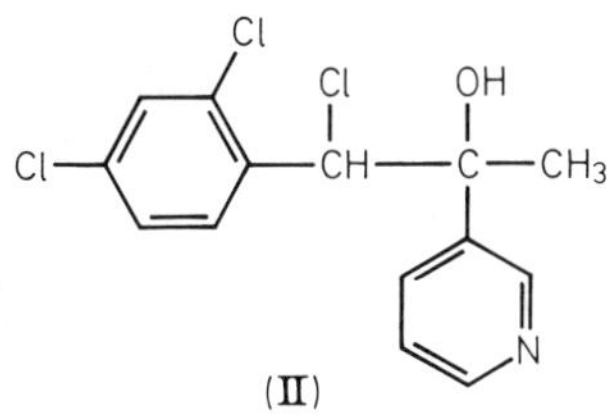

(II)

Pyridinemethanols also have plant growth regulator properties as do some of the pyrimidine analogues of parinol described in the following section. Their use in the control of undesired growth of aquatic plants has been recently described [23, 24, 25, 26].

4 Pyrimidines

The most active fungicides inhibiting sterol biosynthesis [27] come from the pyrimidine class. Ancymidol, the only commercialised plant growth regulator to date whose mechanism of action is based upon inhibition of gibberellic acid biosynthesis at the oxidation step of ent-kaurene to ent-kaurenol, also belongs to this class [27, 27a]. It has already been marketed in many countries.

The commercial products *fenarimol* and *nuarimol* and the experimental product *triarimol*, as well as the other experimental products from this class, can be traced back to a basis patent by Davenport, Hackler and Taylor (Eli Lilly Co.) from 1967 [28] and can be regarded as homologues of the corresponding pyridines.

α-(2-Chlorophenyl)-α-(4-chlorophenyl)-5-pyrimidinemethanol

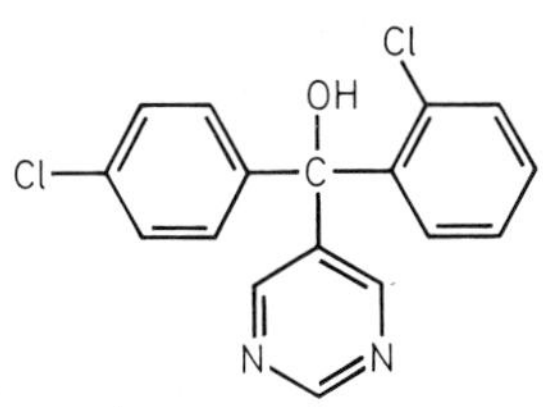

EL 222
Fenarimol, Rubigan®
Eli Lilly
$LD_{50_{rat, oral, acute}}$: 2500 mg/kg

This compound, which is used as a foliar fungicide, has a melting point of 117–119 °C and is soluble in organic solvents such as acetone, acetonitrile, benzene, chloroform

and methanol but only poorly soluble in hexane. Its water solubility is 13.7 ppm at 25 °C and pH 7.0; its vapour pressure is less than 1×10^{-7} mm Hg at 25 °C.
The substance is sensitive towards degradation by UV-light.
α-(2-Chlorophenyl)-α-(4-fluorophenyl)-5-pyrimidinemethanol

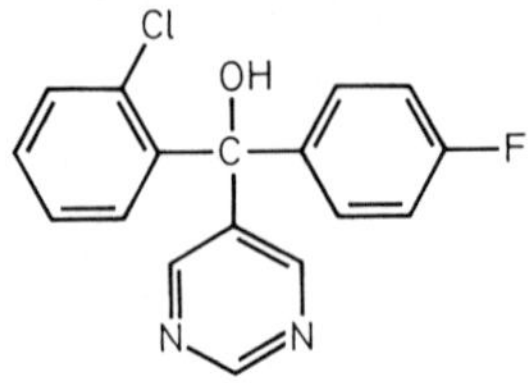

EL 228
Nuarimol, Trimidal®
$LD_{50_{rat,\,oral,\,acute}}$: 2500 mg/kg (female)
1250 mg/kg (male)

Used as a seed treatment, nuarimol has a melting point of 126–127 °C. Its vapour pressure is $<2 \times 10^{-8}$ mbar at 25 °C and it is freely soluble in acetone, acetonitrile, benzene, chloroform and methanol but poorly soluble in hexane. At pH 7.0 and 25 °C 26 ppm dissolves in water. Just like fenarimol, nuarimol is degraded by UV-light.

α-Cyclopropyl-α-(4methoxyphenyl)-5-pyrimidinemethanol

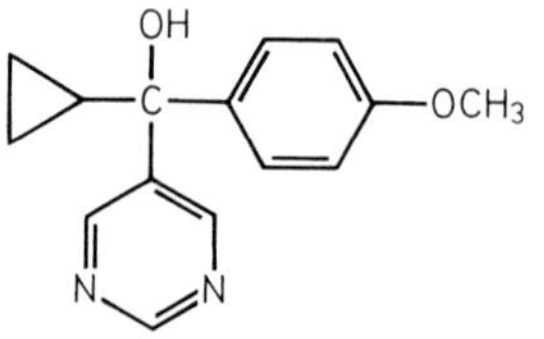

EL 531
Ancymidol, A-Rest®
Reducymol®
Eli Lilly Co.
$LD_{50_{rat,\,oral,\,acute}}$: >2000 mg/kg

The plant growth regulator ancymidol, melting point 110–111 °C, is much more soluble in water (650 ppm) than fenarimol and nuarimol.

It is transported in the phloem.

α-(2,4-Dichlorophenyl)-α-phenyl-5-pyrimidinemethanol

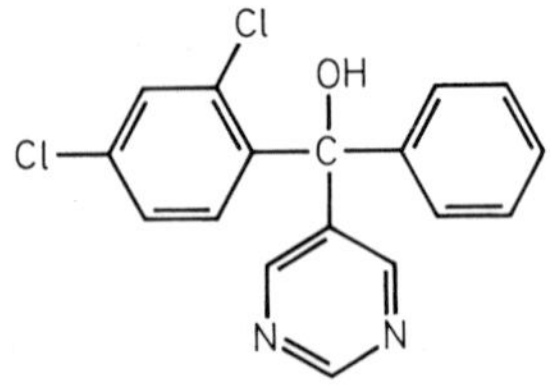

EL 273
Triarimol, Elancocide®
Eli Lilly Co.

This experimental product, melting point 97–98 °C, also attracted considerable attention owing to its systemic activity at very low application rates against powdery mildew and scab fungi [29].
The following experimental products are also worthy of mention.
α-(Chlorophenyl)-α-cyclohexyl-5-pyrimidinemethanol [30]

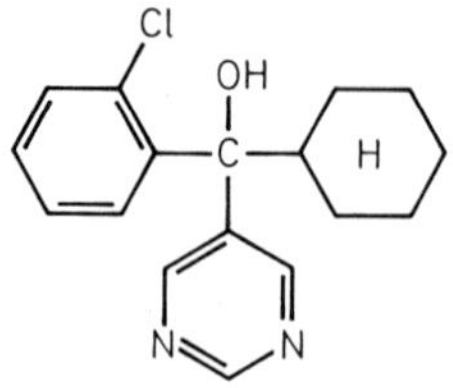

EL 279
systemic fungicide

α-(4-Chlorophenyl)-α-(1-methylethyl)-5-pyrimidinemethanol [27]

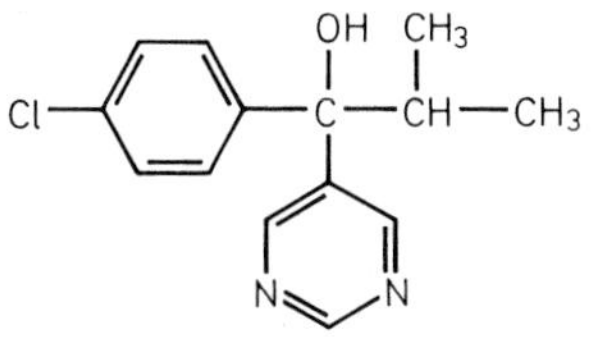

EL 509
plant growth regulator

All these pyrimidines are synthesized from the corresponding ketones by reaction with 5-lithium-pyrimidine (from 5-bromopyrimidine) in THF/Ether at −125 °C as described for ancymidol [31].

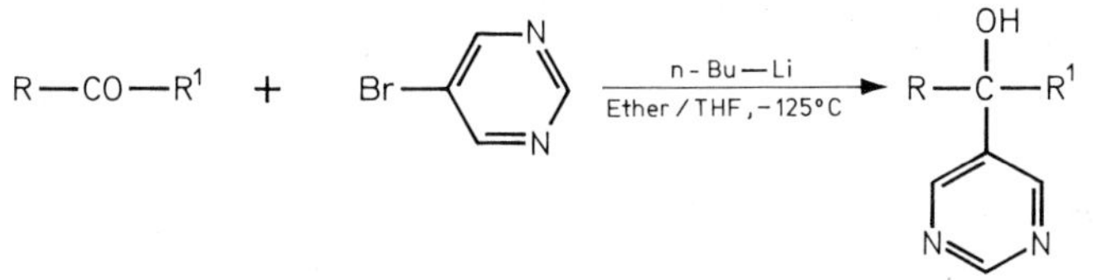

In the case of triarimol the oxidation of the corresponding pyrimidinemethane in DMSO can also be carried out in good yield [21].

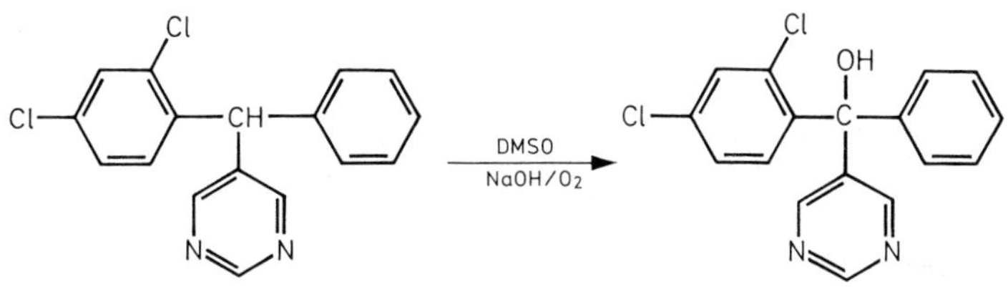

As has been shown in comparative experiments, the fungicidal activity of fenarimol, nuarimol, and triarimol is greater than that of ancymidol and EL 509, the latter two compounds being much more active as growth regulators and gibberellin biosynthesis inhibitors in higher plants [32]. The plant growth regulator activity of the pyrimidinemethanols is only slightly reduced by replacement of the hydroxy group by hydrogen or ethoxy. Replacement of the isopropyl or cyclopropyl group by phenyl or substituted phenyl has a drastic effect on growth regulator activity. A halogen in place of the 4-methoxy group in ancymidol has little effect on plant growth regulator activity; substitution in the 2-position reduces this activity drastically [27]. There is a little difference in lipophilicity between triarimol and fenarimol (log P_{oct} ~ 2.6); nuarimol is markedly more hydrophilic (log P_{oct} ~ 2.0; comparable to that of triadimefon) [33].

5 Antibiotics

Among the antibiotics that interfere in fungal sterol biosynthesis three groups may be differentiated:

Polyene antibiotics of the pimaricin and filipin type
Nucleoside antibiotics of the sinefungine type
Azasteroids of the A 25 822A type.

The products nystatin and amphotericin B, developed as antimycotic drugs, should also be mentioned. The group of antibiotics that interfere in sterol biosynthesis has no technical significance in crop protection at the present time. Only pimaricin is used as an agrochemical for the control of storage rots in strawberries [34] and for the treatment of narcissus bulbs in Holland against Fusarium oxysporum f.sp. narcissi.

5.1 Polyene Macrolides

Most of the polyene antibiotics are produced by Streptomyces species. They have a very large lactone ring (26 ring atoms [pimaricin], 38 [amphotericin B], 44 [lienomycin [35]]) with a lipophilic ring section composed of a polyene chain with 3 to 7 conjugated double bonds and a hydrophilic section with a series of hydroxy groups. The length of the conjugated double bond section permits rapid allocation to the corresponding subgroups by means of their characteristic UV-spectra, but is also responsible for the instability of some polyenes towards heat and light as well as for their sensitivity towards oxidation. These instabilities have limited the practical utility of pimaricin to seed treatment. The fungicidal activity of the polyene macrolides originates from their ability to form complexes with membrane sterols [36, 37].

Pimaricin [38, 39, 40]

Synonym — natamycin
Source: Streptomyces natalensis

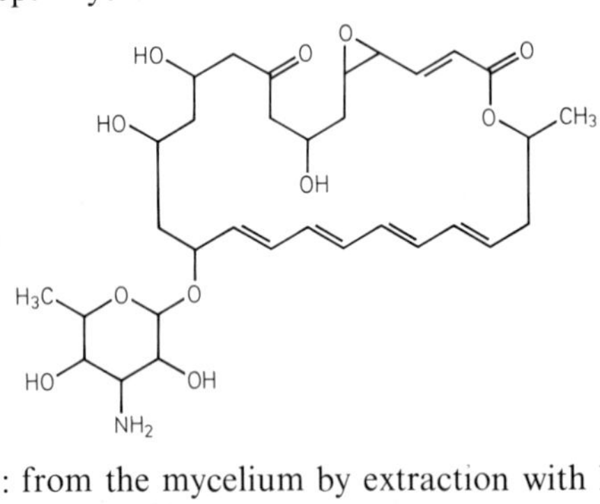

LD_{50} mouse p.o.: 1500 mg/kg

Isolation: from the mycelium by extraction with lower alcohols or formamide.

White crystals with amphoteric character, decomposition point 150 °C, $[\alpha]_D^{25} = 250°$ (c = 0.083% in MeOH).

Studies on its mechanism of action have been carried out by Dekker [41].

Filipin [38, 42]

Synonym — 14-deoxylagosin
Source: Streptomyces filipensis NRRL 2437

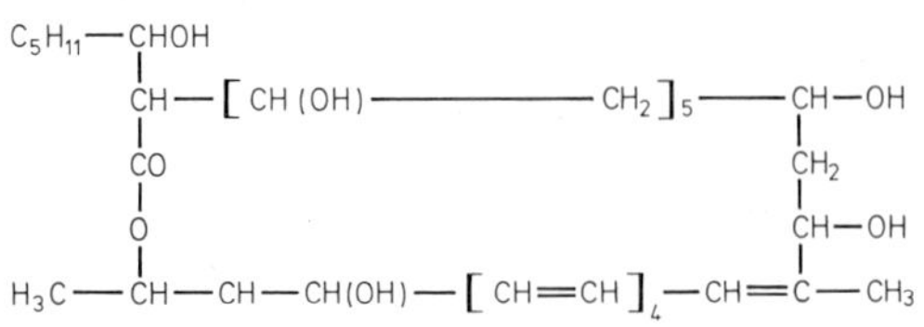

LD_{50} mouse i.p.: 17 mg/kg

Isolation: extraction of the culture filtrate with EtOAc, Me_2CO or BuOH. $[\alpha]_D^{22} = -148.3°$ (c = 0.89%, MeOH); freely soluble: DMF, pyridine, EtOH, MeOH EtOAc; insoluble: water, $CHCl_3$.

Of the antimycotics from the polyene macrolide group, amphotericin B [43] is marketed for the treatment of severe, disseminated mycoses, and nystatin [41] and pimaricin are used as local antimycotics.

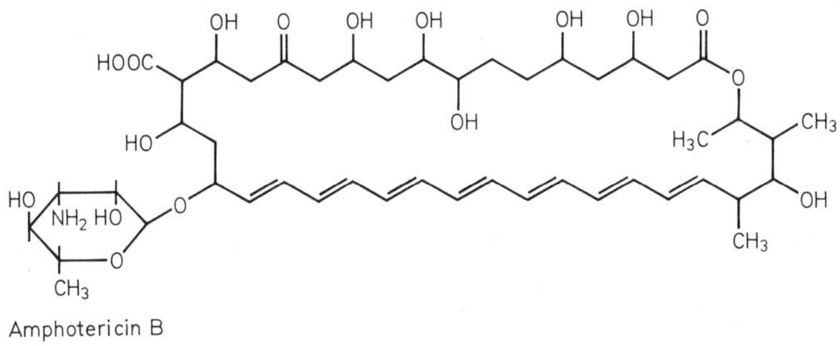

Amphotericin B

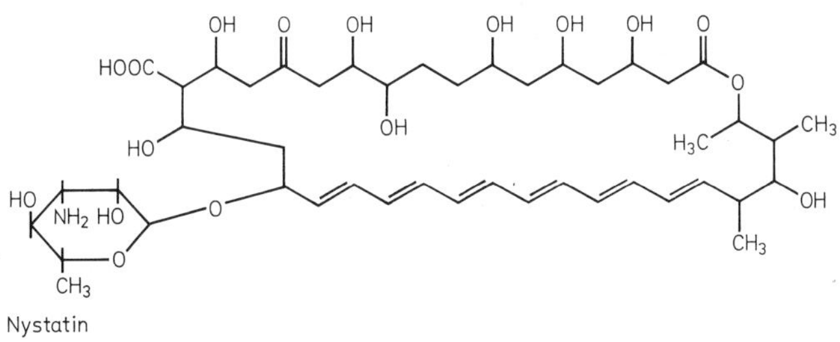

Nystatin

5.2 Nucleoside Antibiotics

As an analogue of S-adenoxylmethionine, sinefungine inhibits the $\triangle^{24}$-methylsterol-c-transferase in *Saccharomyces cerevisiae* [45], causing a build-up of zymasterol.

Sinefungine (A 9145)

Source: *Streptomyces griseolus* NRRL 3739

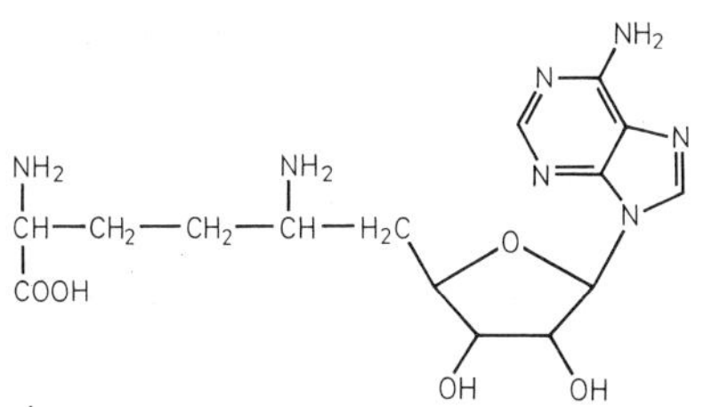

LD_{50} mouse s.c.: 125 mg/kg

Isolation: absorption on acidic JRC-50 ion-exchange resin, elution with 0.05 N H_2SO_4, neutralisation with $Ba(OH)_2$, precipitation with MeOH [46,47].

White amorphous powder, highly soluble in water, slightly soluble in alcohols and most other organic solvents.

As a base it forms salts (HCl-salt, m.p. 195–197 °C).

Sinefungine is comparable to amphotericin B in its activity against *Candida albicans* and is active against phytopathogenic fungi and bacteria such as powdery mildew, bean rust, anthracenosis, *Agrobacterium tumefaciens* and *Xanthomonas phaseoli* var. *sojensis*.

5.3 Azasteroids

Among the azasteroid class 3 substances have been identified that also interfere at various steps in sterol biosynthesis [48].

Antibiotic A 25 822A [49]

Source: *Geotrichum flavobrunneum* NRRL 3862

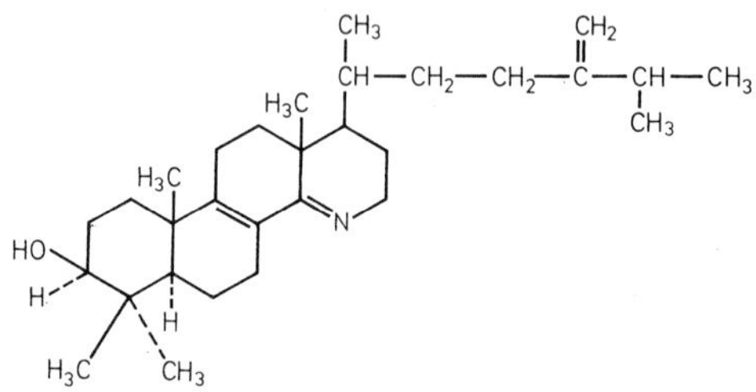

Isolation: extraction with ethyl acetate, 0.2 N H_2SO_4 and, after neutralisation, with hexane. Purification by column chromatography on basic Al_2O_3 with hexane/ethyl acetate/water.

Colourless crystals, m.p. 147 °C, $[\alpha]_D^{25} = -70°$ (c = 1.15, MeOH), pk_a value 8.45 in 66% DMF.

Freely soluble inorganic solvents; not soluble in water.

The strain *Geotrichum flavobrunneum* NRRL 3862 also produces a series of other 3β-hydroxy-24-methylene-14α-aza-D-homo-5α-cholest-8(9),14(14α)dienes such as A 25822B, the 4,4′-desmethyl analogue of A 25822A, its acetate and oxo derivative which have comparable activity against yeasts, dermatophytes, and biphasic fungi, and also against phytopathogenic fungi such as *Botrytis*, *Fusarium* and *Verticillium* sp. as well as against bacteria. The 15-azasterol inhibits ergosterol biosynthesis and causes a build-up of ignosterol[50] in *Saccharomyces cerevisiae*. Detoxification results from S-adenosylmethionine-dependent methylation.

25-Aza-24,25-dihydrozymasterol

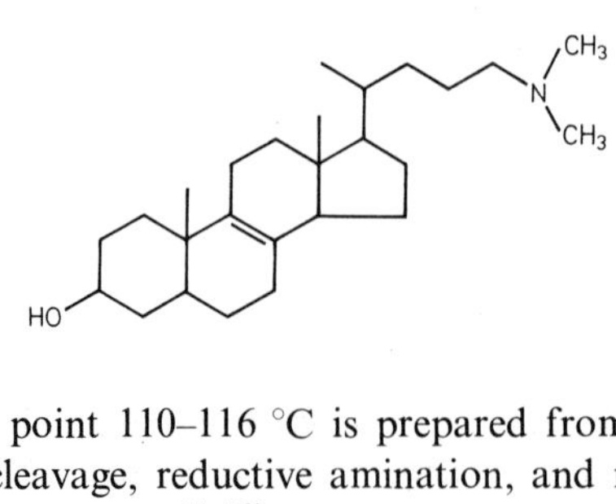

of melting point 110–116 °C is prepared from zymasterol acetate by epoxidation, oxidative cleavage, reductive amination, and reductive ester cleavage according to the following scheme [51,52].

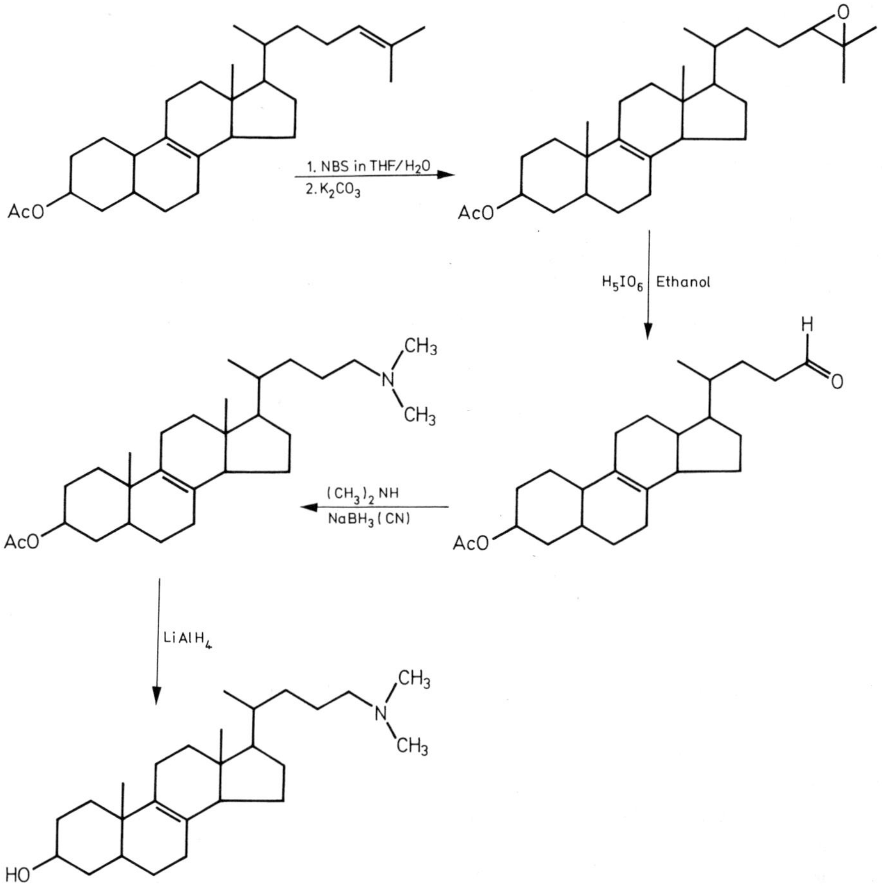

This compound inhibits ergosterol biosynthesis in Saccharomyces cerevisiae by blocking 24-methyl transferase.

23-Azacholesterol

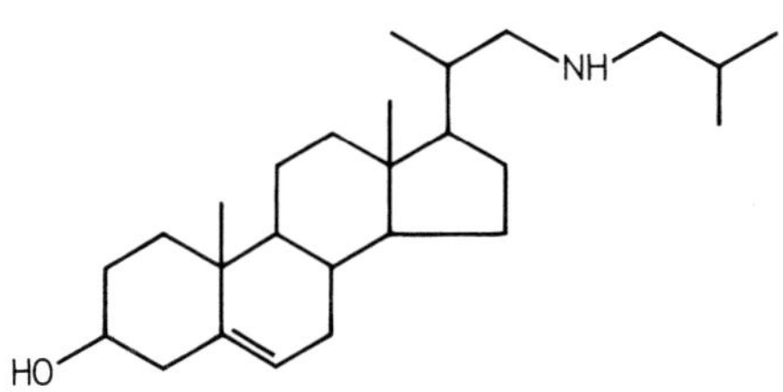

is synthesized in a similar way via 3β-benzoxy-5,7-cholestadiene-22-al [54] starting from stigmasterol. This compound inhibits ergosterol biosynthesis by blocking 24-methylene-24(28)-sterol-reductase [53].

6 Azoles

'Azole fungicides', a group of 1-substituted imidazoles and 1,2,4-triazoles, represent one of the largest classes of fungicides and antimycotics today acting as sterol biosynthesis inhibitors in fungi. The first reports on 1-substituted azole derivatives and their fungicidal properties were made at the end of the sixties by research groups at Bayer and Janssen [55–57]. Büchel and coworkers described the 1-tritylazoles, a class of azole derivatives characterised by high activity against numerous fungal pathogens [58, 59]. Independently the Janssen research group discovered the fungicidal properties of β-alkoxy-β-phenethylimidazoles [57, 60]. The first-ever azole antimycotics (e.g. clotrimazole, miconazole) and azole fungicides (e.g. fluotrimazole, imazalil) to be developed for practical applications stem from these two groups of compounds. In the years following, intensive research was initiated worldwide on imidazole and triazole derivatives after the extraordinarily high potency of these compounds had been recognised. Eleven products from the imidazole and triazole class have since been introduced to the market. These are Triadimefon (Bayleton®), Propiconazole (Tilt®, Radar®), Etaconazole (Sonax®), Triadimenol (Baytan®), Bitertanol (Baycor®), Diclobutrazole (Vigil®), CGA 71 818 (Topas®), Fluotrimazole (Persulon®), Flutriafen (Impact®), Prochloraz (Sportak®), and Fungaflor (Imazalil®). The estimated market turnover of this group in 1983 was $ 220 m. Western Europe accounts for

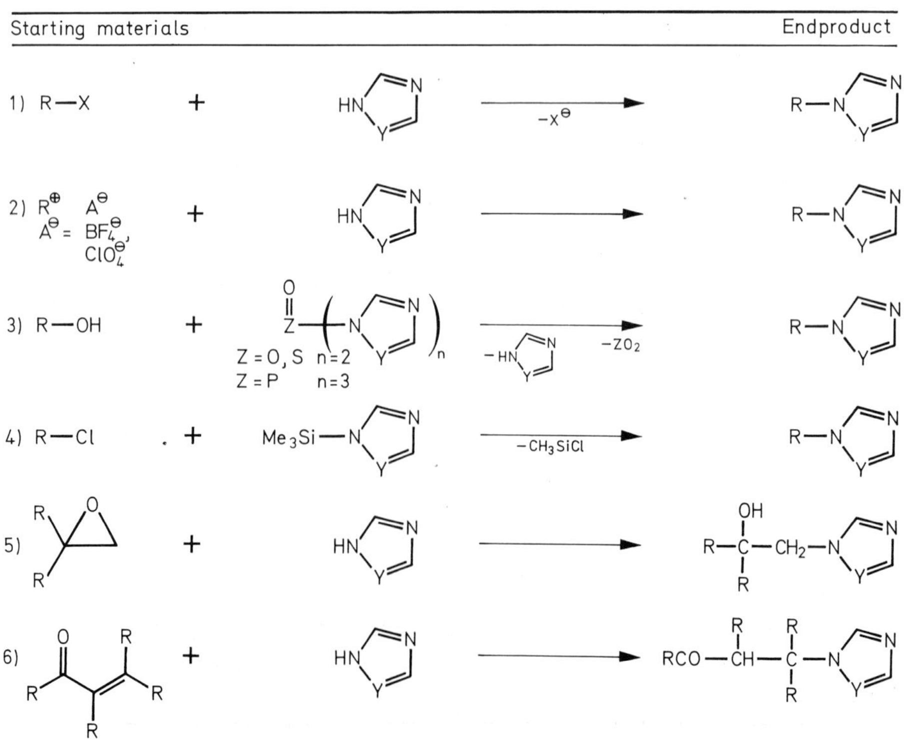

Fig. 1. Synthetic methods for the formation of C-N-bond

approximately 85% of the sales of these fungicides, where the most important outlet is cereals. But the products are also used on top fruits, vegetables and vines. Further compounds are under development from various companies, not only as fungicides but also as plant growth regulators. In the synthesis of N-substituted azole fungicides one of the crucial steps is the creation of the C-N bond between a suitable organic substituent and the azole nitrogen in position 1. Many methods have been worked out for this step, each possessing its own particular value depending on the nature of the residue R.

The process most frequently used is a direct N-alkylation of the azole by a suitably functionalised alkyl derivative, in the presence of base. Alkylation of 1,2,4-triazole (Y = N) leads mainly to the formation of the desired 1-alkylderivative; the corresponding 4-alkyltriazole is formed to a slight to moderate extent depending on the reactivity

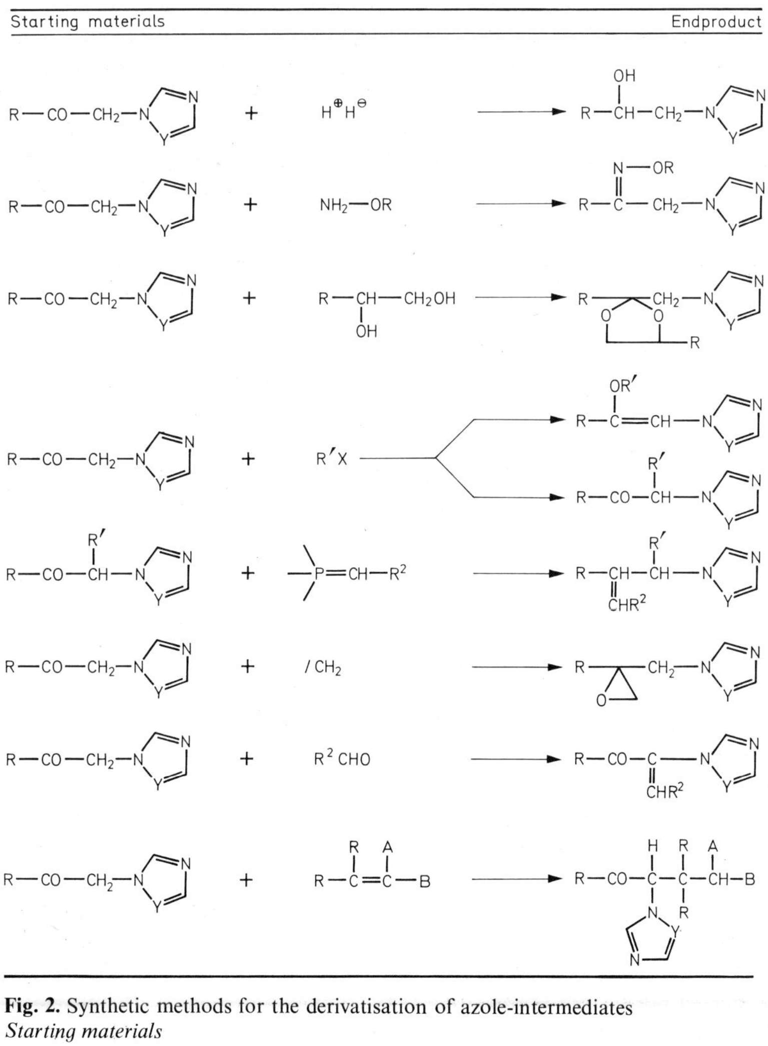

Fig. 2. Synthetic methods for the derivatisation of azole-intermediates
Starting materials

of the alkylating species. In some cases it is possible to rearrange the 4-alkyltriazole into the corresponding 1-alkyltriazole. The use of salts of stable carbocations with nonpolarisable anions is a very special alkylation method applicable in only a few cases, for example, the use of tritylium salts [61]. Azolides of inorganic acids, such as carbonyl- or thionyl-dimidazole, are the reagents of choice in the synthesis of substituted azoles from sensitive alcohols where the corresponding halides either cannot be prepared (e.g. substituted propinols) or are only accessible with difficulty [62].

Another very gentle alkylation method is the cleavage of the Si-N bond of N-trimethylsilyl-azoles with sensitive alkylhalides [63]. The base-catalysed ring opening of oxiranes is a simple route to substituted hydroxy-ethyl-azoles. The addition of azoles to activated double bonds yields the corresponding N-alkylazoles (Michael adducts). However, this method is limited by the reversibility of the reaction [64,65].

Another method for the synthesis of azole fungicides is the derivatisation of suitable 1-substituted azole intermediates. For example, 1-substituted azolyl-methylene-ketones, 1-substituted hydroxy-ethyl-azoles and halogenalkyl-azoles are such suitable compounds. In the case of 1-substituted azolyl-methylene-ketones derivatisation reactions are, for example, reduction, oximation, ketalisation and epoxidation of the keto group, alkylation at the methylene or oxofunction [66,67,68], 'Wittig-reaction' at the oxo function [69,70], aldol-condensation [71] and Michael addition due to the C-H acidity of the methylene-function [72]. An overview is given in Fig. 2.

Structure activity correlations in the field of azoles show that the essential parameters for good activity, apart from the suitable lipophilicity of the compounds, are stereochemical parameters [73] and, in general, a suitable stereochemistry of the compounds due to the fact that the azoles bind with their lipophilic part of the molecules to the lipophilic binding site of cytochrome P_{450}, which is normally occupied by the sterol precursors of ergosterol biosynthesis in fungi [74]. Even though the stereochemistry of azole fungicides and the shape of the active molecules are the most important factors for their activity and limit the finding of active compounds, the wide variability of the substitutent R in the 1-position of the imidazole or 1,2,4-triazole residue (without loss of antifungal activity) initiated worldwide research in the field of azole fungicides and antimycotics. This variability will be illustrated by the compounds described in the following.

6.1 Azoles Without Other Functional Groups

Within this highly variable group of azoles, two products were developed as fungicides. One of them, fluotrimazole belongs to the N-tritylazole group, the first azole group discovered by Büchel and coworkers at Bayer [55–57]. The second, CGA 71818 (Topas®), from the group of 1-(2-arylethyl)-1H-1,2,4-triazoles, was found by Heeres and coworkers in 1976 [75], developed by Ciba Geigy and introduced into the market in 1983.

Important antimycotics from this group are clotrimazole (Canesten®), and bifonazole (Mycospor®) from Bayer AG. Clotrimazole was the first azole antimycotic, launched in 1972; bifonazole was introduced in 1983. As an anti-acne compound the 2-chloro-analogue of bifonazole, lombazole, is the active ingredient of the cosmetic lotion Twent®.

Diphenyl-(3-fluoromethylphenyl)-1,2,4-triazol-1-yl-methane [76]

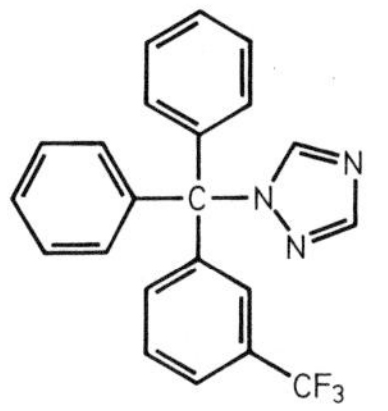

Fluotrimazole
Persulon®
Bayer AG
$LD_{50_{\text{rat, oral, acute}}}$: > 5000 mg/kg

This compound, a cereal fungicide for control of powdery mildew, was the first azole fungicide to be marketed.

The key-product for its synthesis, m-trifluoromethyl-trityl chloride, is accessible, starting from meta-xylene, by peroxide-catalysed perchlorination and selective chlorine/fluorine exchange [77], or, on a technical scale, by conproportionation of m-trichloromethyl-benzotrichloride and m-trifluoromethyl-benzotrifluoride [78] followed by Friedel-Crafts reaction with benzene. The trifluoromethyltrityl chloride reacts smoothly with 1,2,4-triazole in polar solvents to give the trityl-triazole.

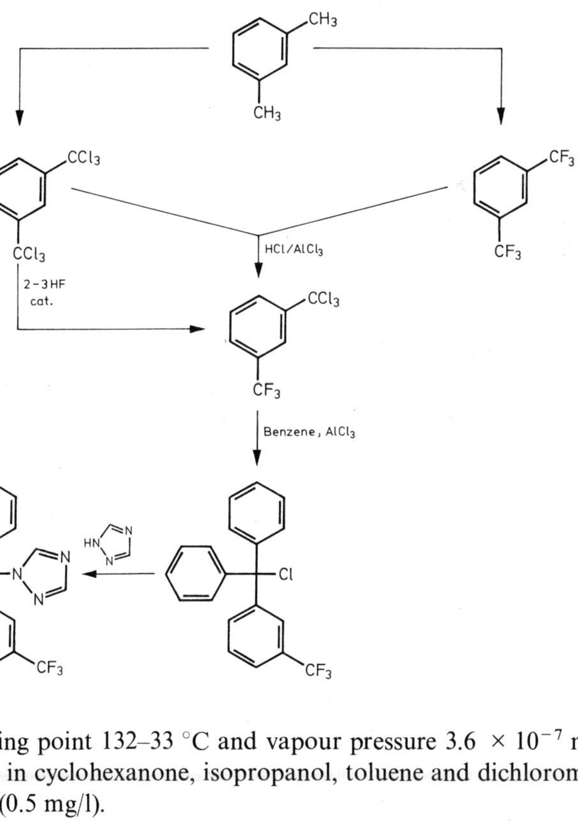

Fluotrimazole, melting point 132–33 °C and vapour pressure 3.6×10^{-7} mbar at 20 °C, is freely soluble in cyclohexanone, isopropanol, toluene and dichloromethane but insoluble in water (0.5 mg/l).

2-(2,4-Dichlorophenyl)-1-(1,2,4-triazol-1-yl)-pentane [75]

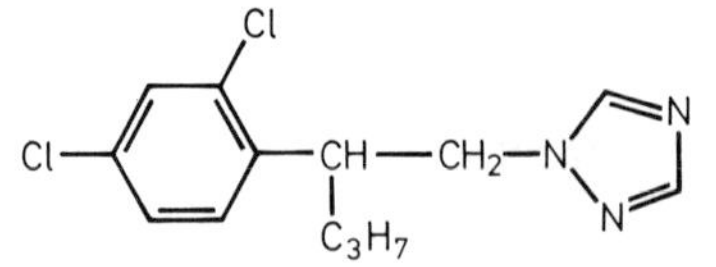

CGA 71818
Topas®
Ciba Geigy
$LD_{50_{rat, oral, acute}}$: 2125 mg/kg

The compound is soluble in most organic solvents, has a vapour pressure of 2.1 × 10^{-6} mbar at 20 °C, and a melting point of 60 °C.

Two different synthetic routes lead to the product. Starting from 2,4-dichlorophenylacetonitrile or 2,4-dichlorophenylacetate, alkylation with propyl halogenide, reduction with lithium aluminium hydride and esterification by sulphonic acid gives the key intermediate, 2-(2,4-dichlorophenyl)-pentanol methane-sulphonate, which reacts with sodium triazolide to give the end-product.

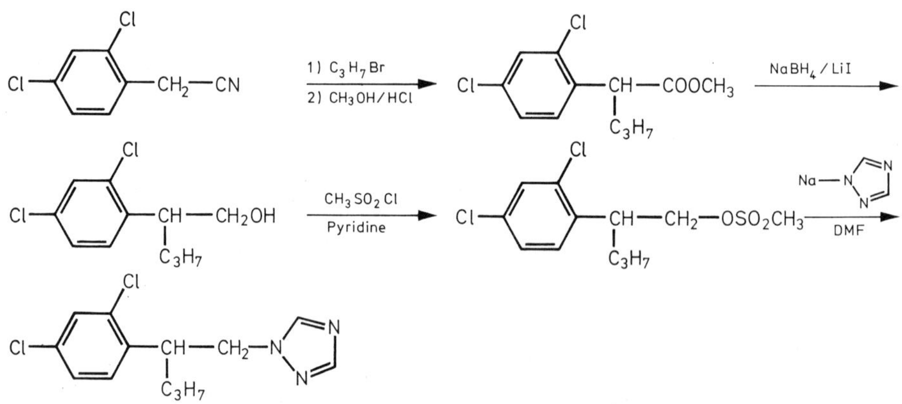

The other route starts from 2,4-dichlorobutyrophenone. The methylene-azole moiety is introduced by a Wittig-Horner reaction with triazolylmethyl-phosphonate. Hydrogenation then gives the desired product [79].

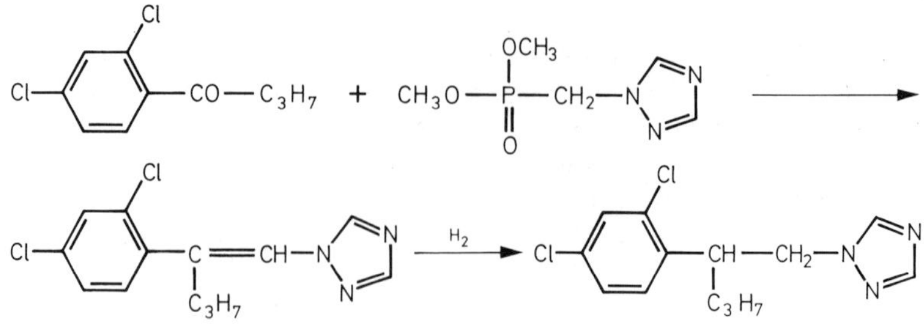

The variability of such compounds which are highly active is also shown by the following antimicrobials.

Table 1. Antimicrobials from azoles without any other functional groups

Compound	Name	Use	Ref.
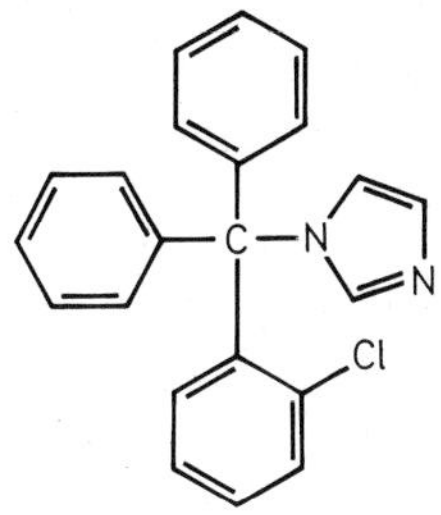	Clotrimazole Canesten®	Local Antimycotic	[56]
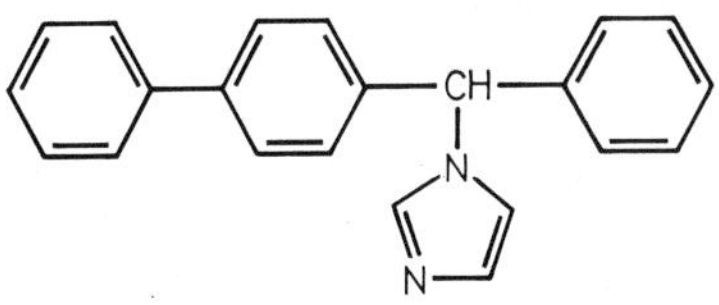	Bifonazole Mycospor®	Local Antimycotic	[80,82]
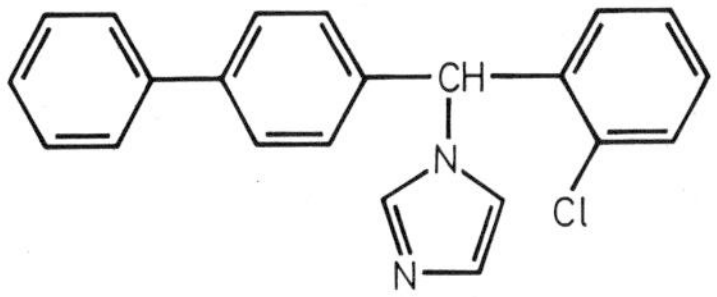	Lombazole Twent®	Antiacne Compound	[80]
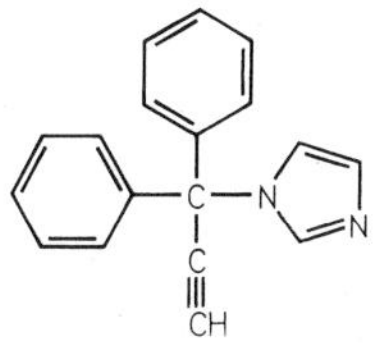	BAY d 9603	Against mould fungi	[81]

6.2 Azoles with Keto Groups

Triadimefon (Bayleton®), BASF LAB 117682, BASF LAB 130827 and climbazole (Baysan®, Baypival®, Ceox®) belong to this azole class. These products also illustrate in an almost classic fashion the breadth of biological activity of the azoles, covering systemic, broad spectrum fungicides, growth regulators and fungicides for household use and antidandruff shampoos.

1-(4-Chlorophenoxy)-3,3-dimethyl-1-(1,2,4-triazol-1-yl)-2-butanone [83]

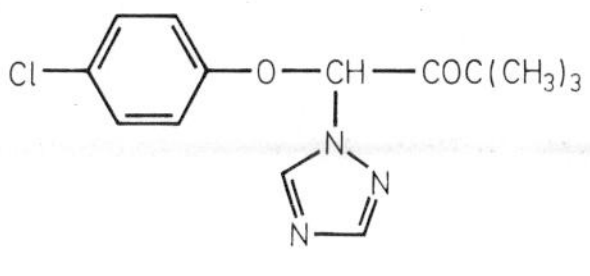

Triadimefon
Bayleton®
Bayer AG
$LD_{50_{rat,\,oral,\,acute}}$: 750–1200 mg/kg

Triadimefon has a melting point of 82.3 °C, a vapour pressure of $<10^{-6}$ mbar at 20 °C, and is freely soluble in cyclohexanone, isopropanol, dichloromethane and toluene but sparingly soluble in ligroin. Its solubility in water is 70 mg/l at 20 °C.

Triadimefon can be synthesized by three different routes starting from pinacolone. Monoclorination, nucleophilic substitution with p-chlorophenoxide, halogenation and nucleophilic substitution with 1,2,4-triazole gives the end product [83].

Alternatively, pinacolone is dichlorinated and then reacted with sodium p-chlorophenoxide and triazole [84] in a one-pot reaction as in the following scheme:

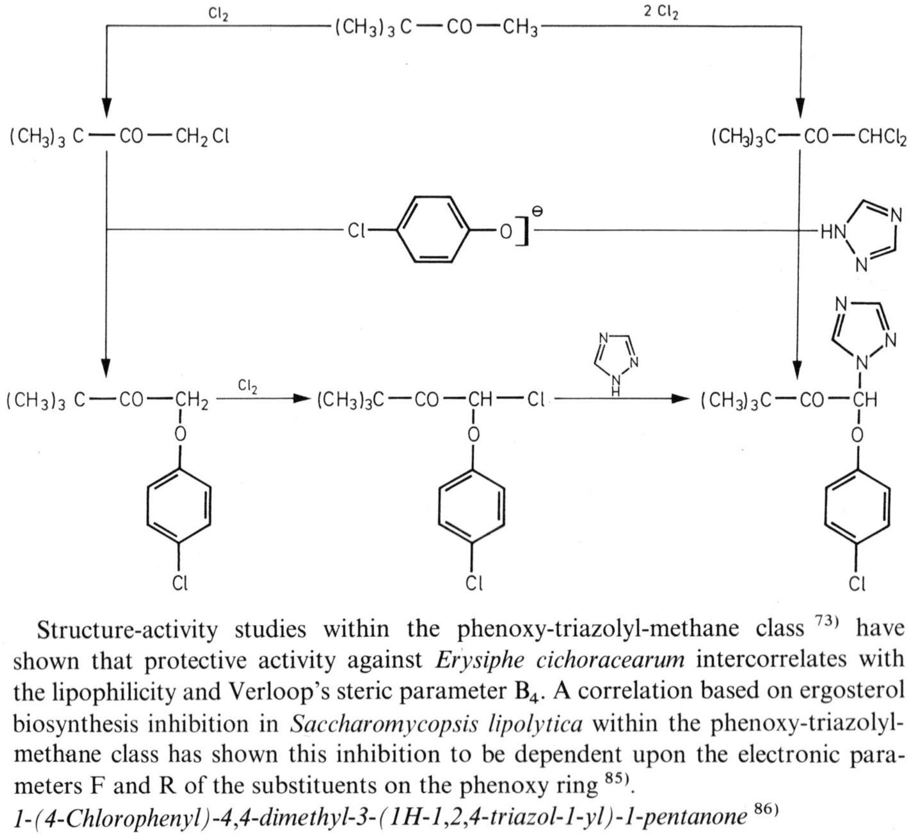

Structure-activity studies within the phenoxy-triazolyl-methane class [73] have shown that protective activity against *Erysiphe cichoracearum* intercorrelates with the lipophilicity and Verloop's steric parameter B_4. A correlation based on ergosterol biosynthesis inhibition in *Saccharomycopsis lipolytica* within the phenoxy-triazolyl-methane class has shown this inhibition to be dependent upon the electronic parameters F and R of the substituents on the phenoxy ring [85].

1-(4-Chlorophenyl)-4,4-dimethyl-3-(1H-1,2,4-triazol-1-yl)-1-pentanone [86]

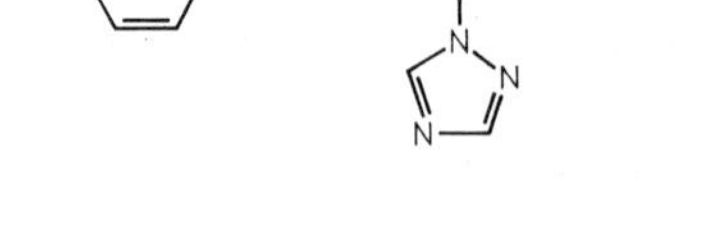

BASF LAB 117 682
m.p. 113–115 °C

and

1-(4-Chlorophenyl)-2,4,4-trimethyl-3-(1H-1,2,4-triazol-1-yl)-1-pentanone

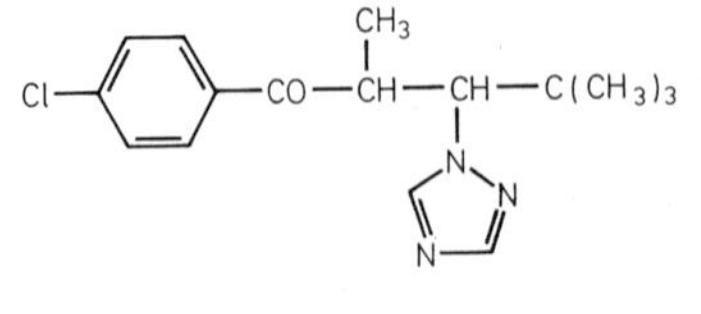

BASF LAB 130 827
$n_D^{20} = 1{,}5172$

have been reported as plant growth regulators [87]. Of these two experimental products from BASF, LAB 130827 is apparently the more active. They are synthesized by Michael addition of triazole to the corresponding 2,3-penten-1-ones.

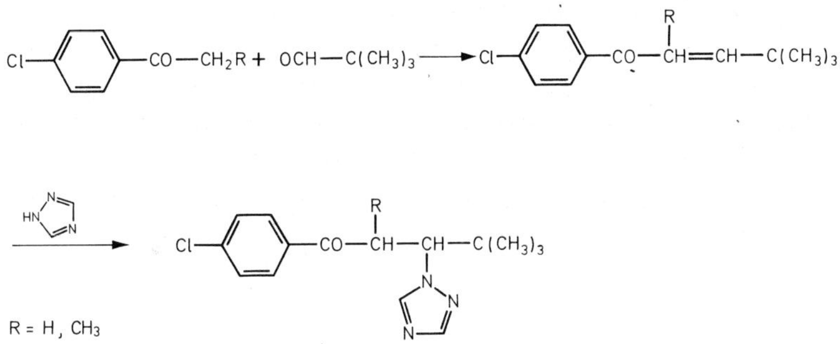

The imidazole analogue of triadimefon,
1-(4-chlorophenoxy)-1-(1H-imidazol-1-yl)-3,3-dimethylbutanone

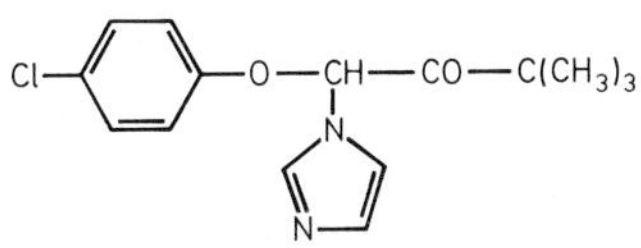

Climbazole
Baysan®, Baypival®, Ceox®
Bayer AG
$LD_{50_{\text{rat, oral, acute}}}$: 400 mg/kg

can be used against *Aspergillus*, *Penicillium*, *Candida* and *Poecilomyces* in the household. As active ingredient of the antidandruff shampoo Ceox® it acts against *Pitorosporum ovale* and thus prevents the formation of dandruff.

The following keto-alcohol is an experimental product from ICI.
(5RS,6RS)-6-Hydroxy-2,2,7,7-tetramethyl-5-(1,2,4-triazol-1-yl)-octan-3-one [88]

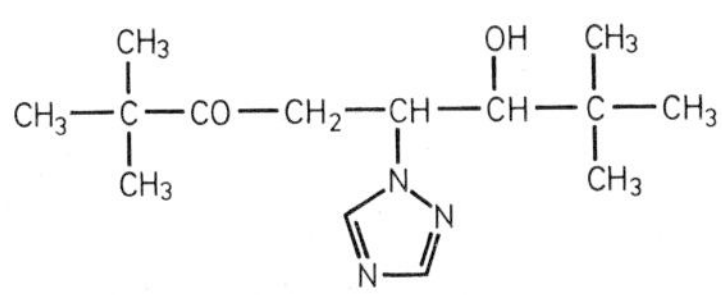

PP 969
ICI

is a highly systemic fungicide active against coffee rust, Sikatoka disease of bananas, apple mildew and apple scab. The substance is taken up via the soil or injected into the tree trunk; uptake and activity via the foliage is lower [89].

This compound is also synthesized via a Michael addition and partial reduction according to the following scheme:

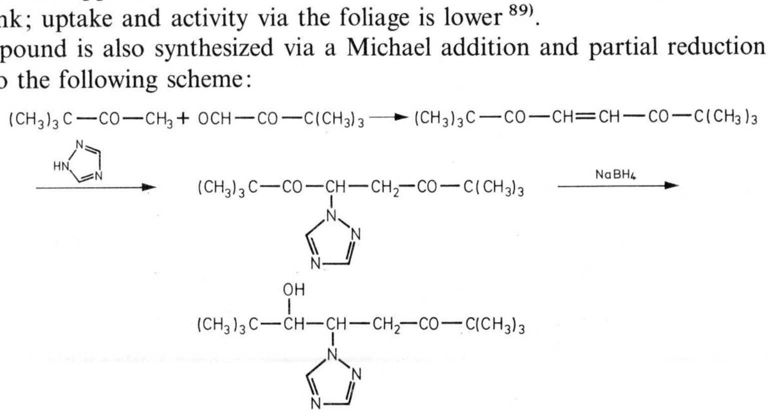

6.3 Azoles with Ketal Groups

This group of azoles includes a series of highly active fungicides and antimycotics mainly discovered in the research laboratories of Janssen Pharmaceutical Co., some compounds being developed by Ciba -Geigy AG.

1-[2-(2,4-Dichlorophenyl)-4-propyl-1,3-dioxolan-2yl-methyl]-1H-1,2,4-triazole [90,91]

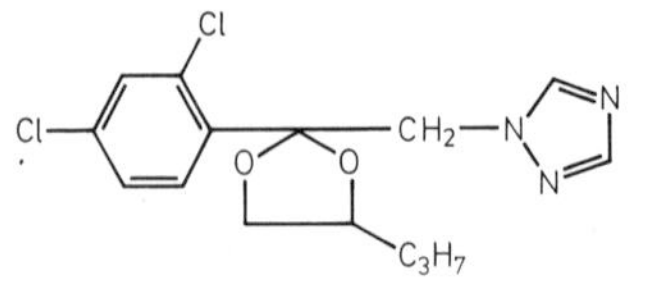

CGA 64250
Propiconazole (proposed)
Tilt® Ciba Geigy AG
Radar® ICI
$LD_{50_{rat, oral, acute}}$: 1517 mg/kg

Propiconazole, marketed as a cereal fungicide, is a colourless liquid with boiling point 180 °C/0.13 mbar, vapour pressure 1.3×10^{-6} mbar at 20 °C, has a water solubility of 110 mg/l at 20 °C and is soluble in most organic solvents such as acetone, toluene and dichloromethane.

Propiconazole is available by a two-step process [90] starting with ω-bromo-2,4-dichloroacetophenone and involving acid-catalysed ketalisation with 1,2-pentanediol then nucleophilic substitution with sodium 1,2,4-triazolide in DMF at 160 °C.

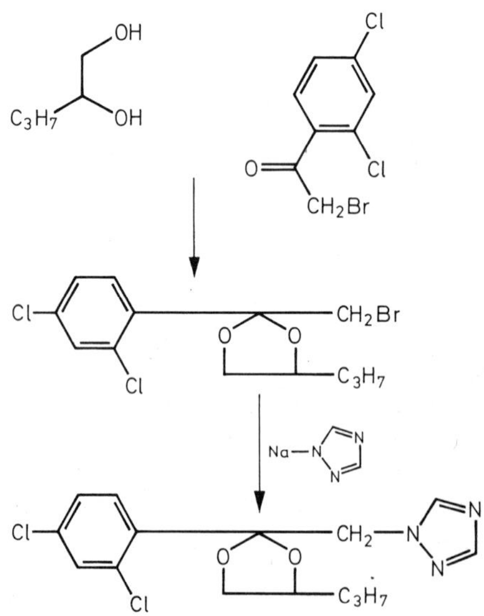

The experimental product,
1-[2-(2,4-dichlorophenyl)-4-ethyl-1,3-dioxolan-2-yl-methyl]-1H-1,2,4-triazole [90,92]

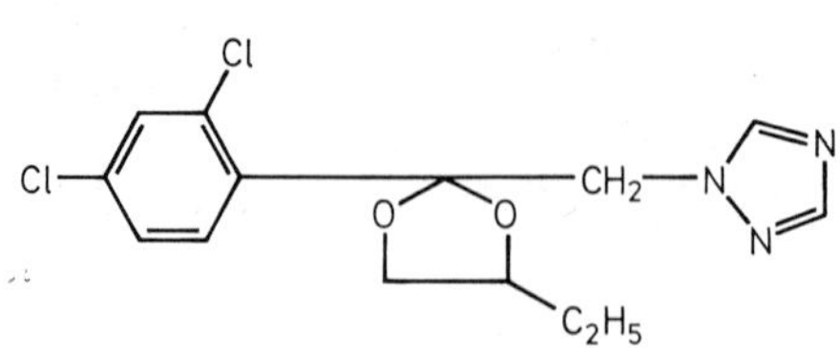

CGA 64 251
Etaconazole
Sonax®
Vangard®
Ciba Geigy AG
$LD_{50_{rat, oral, acute}}$: 1343 mg/kg

is synthesized by the analogous route and currently undergoing trials as a systemic fungicide for the control of diseases of ornamentals, fruits and vegetables.

Structure-activity studies have shown the activity of these azole ketals to be highly dependent upon the chain length of the alkyl group on the dioxolane ring.

Ketoconazole,
cis-1-acetyl-4-[4-[[(2,4-dichlorophenyl)-2-(1H-imidazol-1-yl-methyl)-1,3-dioxolan-4-yl]methoxy]phenyl]piperazine [93)]

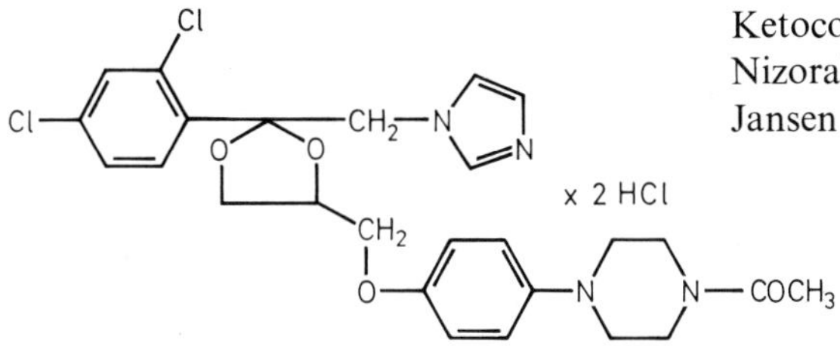

Ketoconazole
Nizoral®
Jansen Pharmaceutical Co.

was the first oral azole antimycotic to be introduced for the treatment of systemic mycoses. Several other derivatives from the dioxolane group have been reported as antimycotics.

Further derivatives of ketone azoles to be reported are:
2,4-Dichloro-ω-(1,2,4-triazol-1-yl)-benzyloximino-acetophenone [98)]

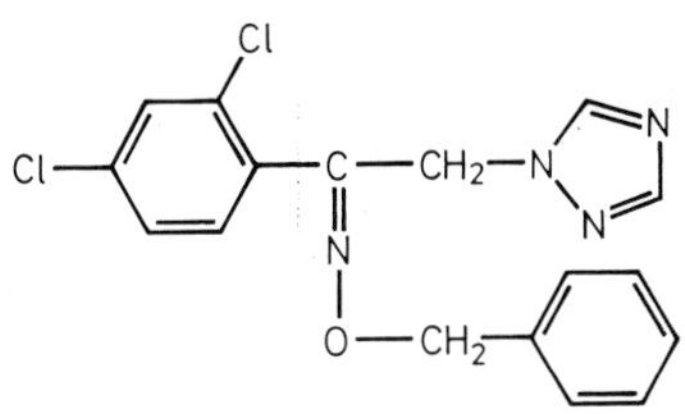

BASF 45 406 F
BASF AG
$LD_{50_{rat,\ oral,\ acute}}$: 4640 mg/kg

This compound, an experimental product from BASF AG is active as a fungicide against Cercospora beticola, Venturia inaequalis, Hemileia vastrix, Septoria nodorum and Erysiphe graminis.

Oxiconazole, a local antimycotic from Siegfried AG, was launched in 1983:
2,4-Dichlorophenyl-ω-(1H-imidazol-1-yl)-2,4-dichlorobenzyl-2-oximino-aceto-phenone [99].

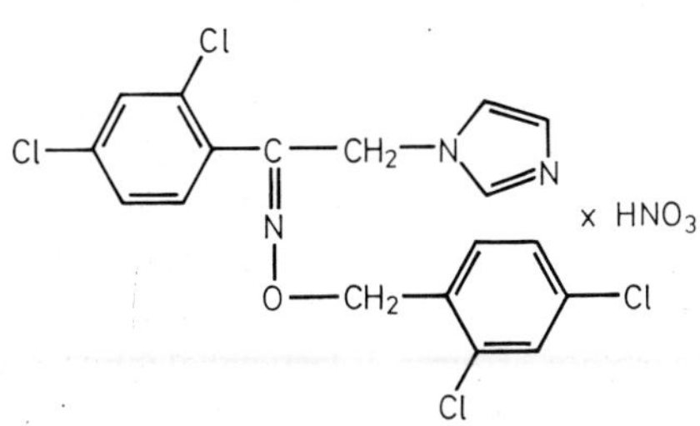

RÒ-13-8996
Oxiconazole
Oceral®
Siegfried AG

Table 2. Antimycotics of the dixolanemethyl-azole group

Structure	Use	Manufacturer	Literature
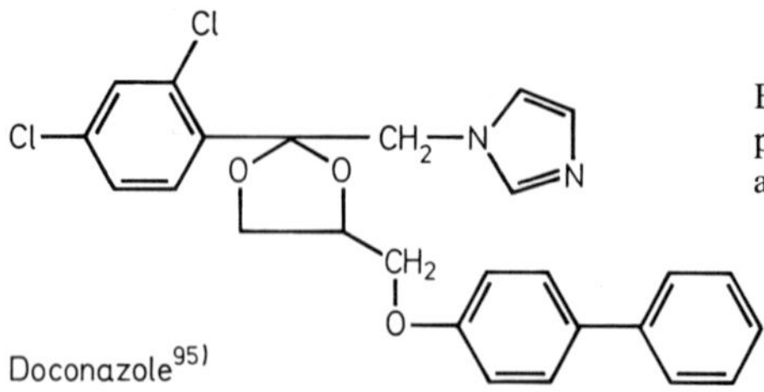 Doconazole[95]	Experimental product oral antimycotic	Janssen	DOS 2602770
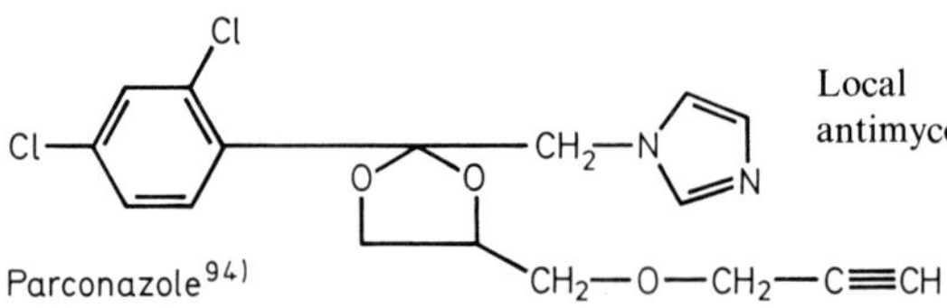 Parconazole[94]	Local antimycotic	Janssen	DOS 2602770 (1975)
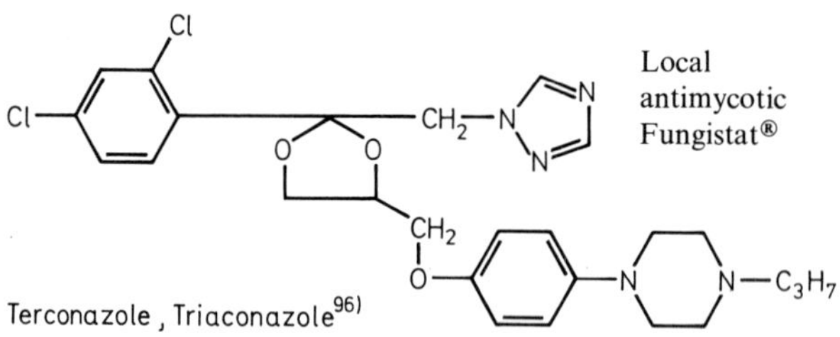 Terconazole, Triaconazole[96]	Local antimycotic Fungistat®	Janssen	DOS 2804096 (1977)
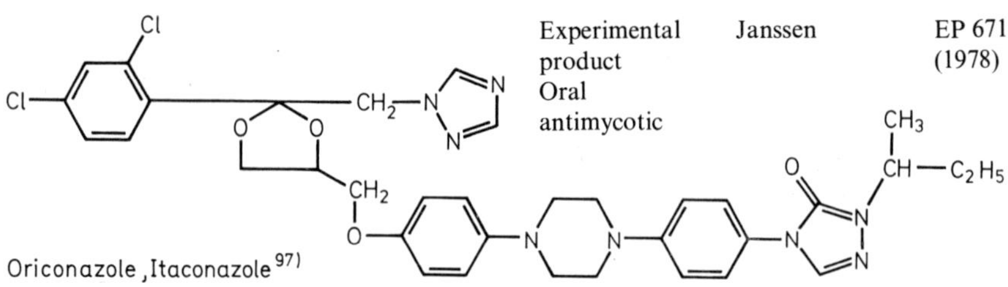 Oriconazole, Itaconazole[97]	Experimental product Oral antimycotic	Janssen	EP 6711 (1978)

6.4 Azoles with Hydroxy Groups

This subgroup of the azoles has produced the most interesting fungicides and plant growth regulators to date. Two types can be differentiated: hydroxy azoles in which the carbon atom substituted by the azole ring also bears further substituents (e.g. triadimenol, bitertanol, pentefenzol (proposed), dichlobutrazole, paclobutrazole, S 3308, RSW 0411), and hydroxy azoles in which the carbon atom bearing the azole ring is only substituted by hydrogen but the carbinol group is a tertiary alcohol (e.g. fluotriafol (proposed)).

β-(4-Chlorophenoxy)-α-(1,1-dimethylethyl)-1H-1,2,4-triazole-1-ethanol [100)]

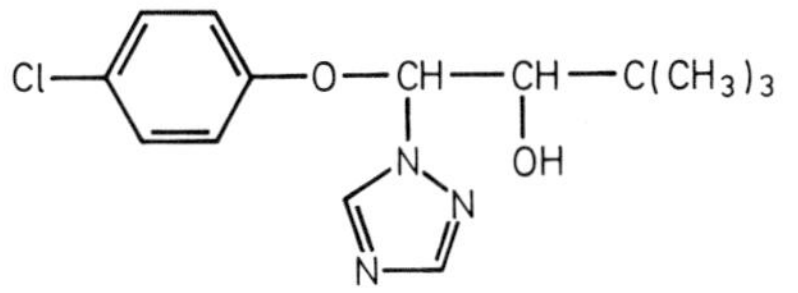

Triadimenol
Baytan®
Bayfidan®
Bayer AG
$LD_{50_{rat, oral, acute}}$: 700–1500 mg/kg

Triadimenol, obtained from triadimefon by reduction with aluminium isopropoxide [101)], has a very slight water solubility (95 mg/l H_2O), is freely soluble in dichloromethane, propanol and cyclohexane, moderately soluble in toluene and sparingly soluble in ligroin.

Vapour pressure $\ll 10^{-6}$ mbar at 20 °C.

Its high systemicity permits use both as a seed treatment and as a cereal spray.

The 4-phenyl analogue, a foliar fungicide and cereal seed treatment,
β-[(1,1'-biphenyl)-4-yloxy]-α-(1,1-dimethylethyl)-1H-1,2,4-triazole-1-ethanol [102)]

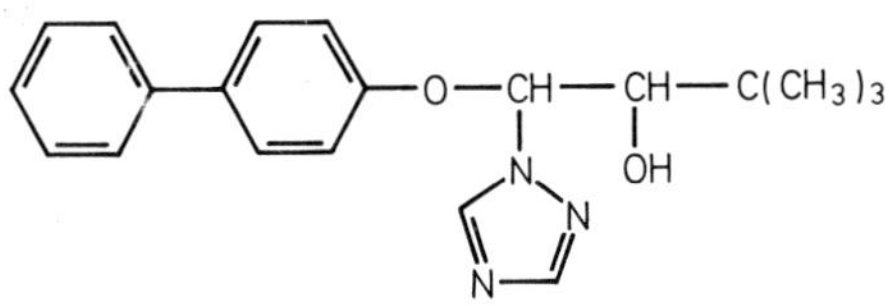

Bitertanol
Baycor®, Baymat-Spray®
Sibutol®
Bayer AG
$LD_{50_{rat, oral, acute}}$: > 5000 mg/kg

is, owing to its increased lipophilicity compared with triadimenol, less systemic but nevertheless locosystemic and curative and has a much lower water solubility of 5 mg/l H_2O. It is soluble in cyclohexane and dichloromethane but only moderately soluble in ligroin, propanol and toluene. Vapour pressure $\ll 10^{-6}$ mbar at 20 °C.

Replacement of the phenoxy group in triadimenol by the benzyl group also leads to highly active fungicides:

β-[(2,4-Dichlorophenyl)-methyl]-2-(1,1-dimethyl)-1H-1,2,4-triazole-1-ethanol [103)]

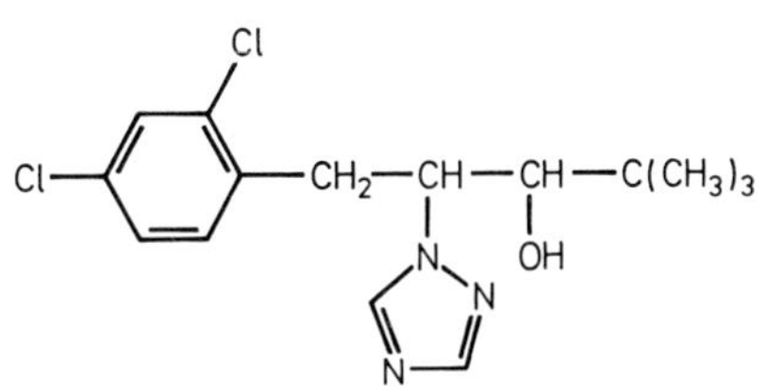

Dichlobutrazole
Vigil®
ICI
$LD_{50_{rat, oral, acute}}$: ca. 4000 mg/kg

Dichlobutrazole, is prepared by condensation of pinacolone with benzaldehyde then reduction, bromination and nucleophilic substitution then sodium borohydride reduction according to the following scheme:

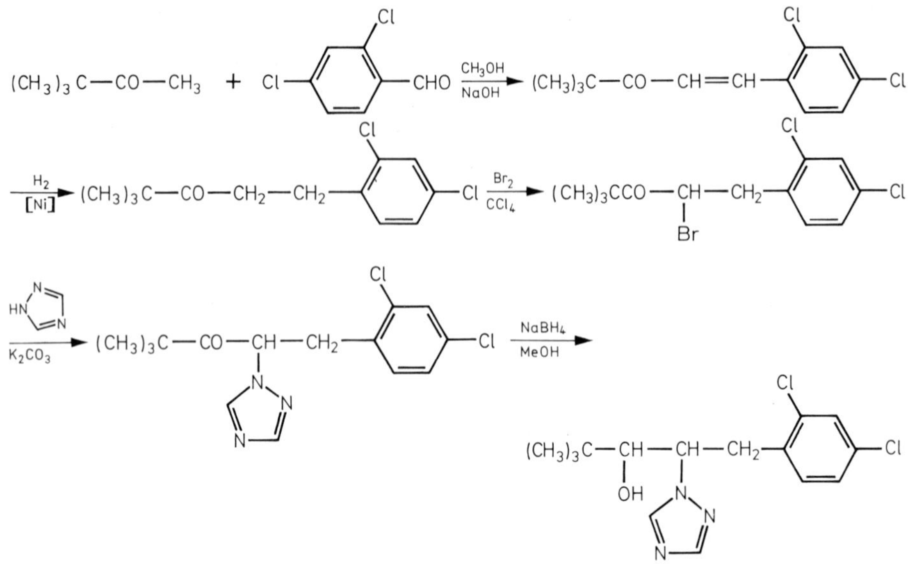

The main product is the RR,SS-diastereomer. An alternative synthesis starts from triazolylpinacolone. Alkylation with 2,4-dichlorobenzyl chloride and subsequent reduction leads to dichlobutrazole:

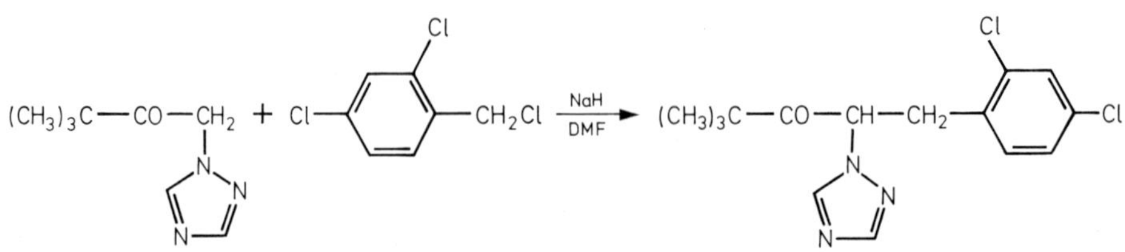

Replacement of the benzyl function by the benzylidene function also leads to highly active fungicides, as illustrated by the development product S 3308 (Sumitomo):

β-[2,4-Dichlorophenyl)-methylene]-α(1,1-dimethylethyl)-1H-1,2,4-triazole-1-ethanol [104]

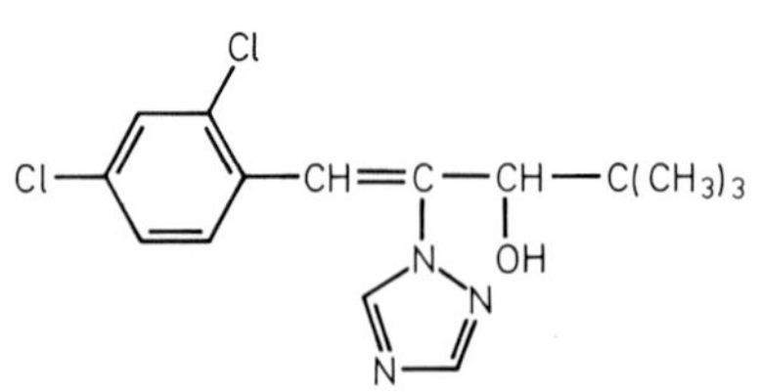

S 3308 Sumitomo

In this group the E-isomers are the more active fungicides, the Z-isomers being less active. These compounds are accessible by condensation of 2,4-dichlorobenzaldehyde with triazolylpinacolone, photochemical rearrangement of the E/Z-mixture formed into the E-isomer and then borohydride reduction:

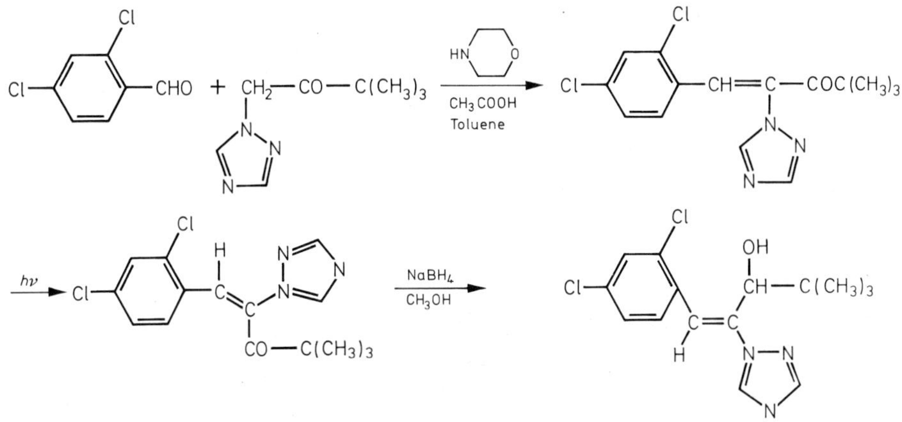

Several compounds from this hydroxy-azole class are under development as plant growth regulators as shown in the following Table:

Table 3. Growth regulators from the triazole-ethanol class

Structure	Designation	Manufacturer	Literature
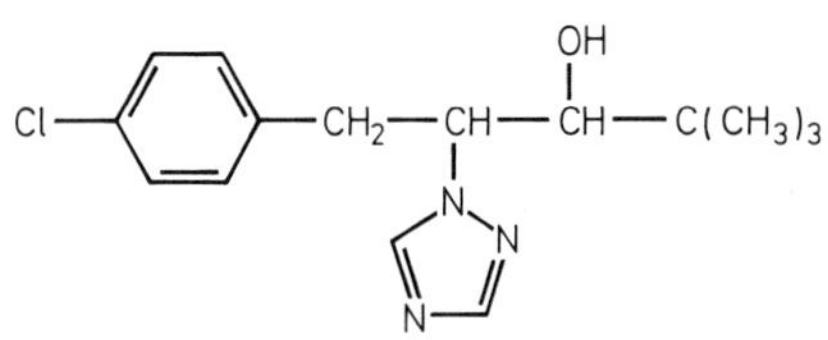	PP 333 Paclobutrazole	ICI	DOS 2737489 (1976)
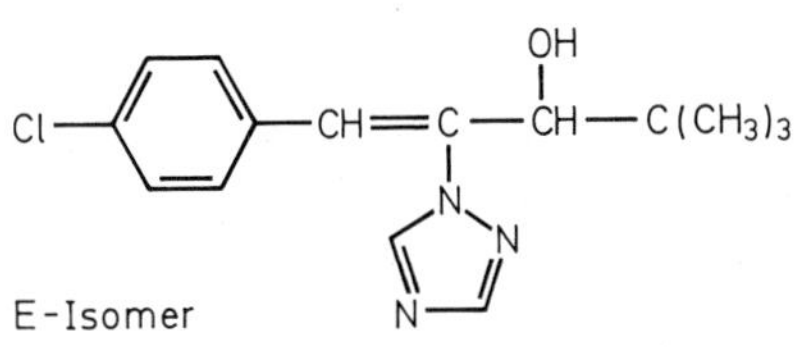 E-Isomer	Pentefenazole Sumiseven® Uniconazole®	Sumitomo	DOS 3010560 (1979)
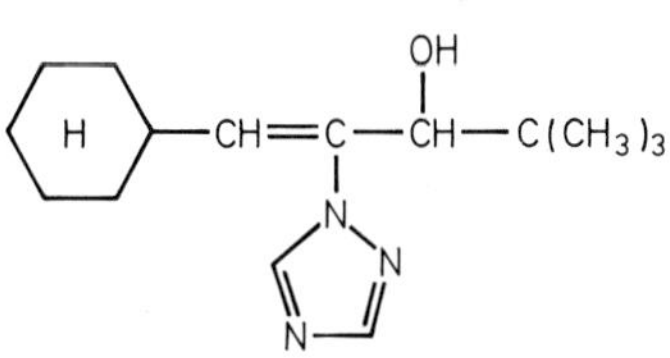	RSW 0411 Triapenthenol (proposed)	Bayer AG	EP 15387 (1979)

The first representative of the tertiary triazole-alcohols to reach the market is PP 450: *(RS)-2,4'-Difluoro-α-(1H-1,2,4-triazol-1-yl-methyl)-benzhydrylalcohol* [105]

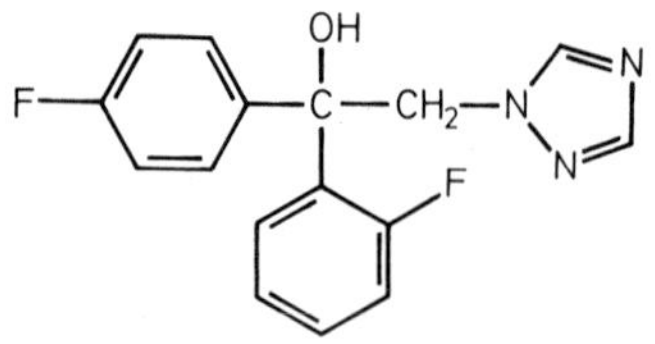

PP 450
Flutriafen, Fluotriafol
(proposed)
Impact®
ICI
$LD_{50_{rat, oral, acute}}$: 1140 mg/kg (male)
1480 mg/kg (female)

PP 450 has a melting point of 130 °C and a vapour pressure of 4×10^{-9} mbar at 20 °C. It is freely soluble in acetone, dichloromethane, methanol and xylene, only slightly soluble in hexane, and has a water solubility of 0.13 g/l at pH 7.0.

For the synthesis the corresponding benzophenone is converted to the epoxide by the Corey method and the epoxide opened by triazole [106]

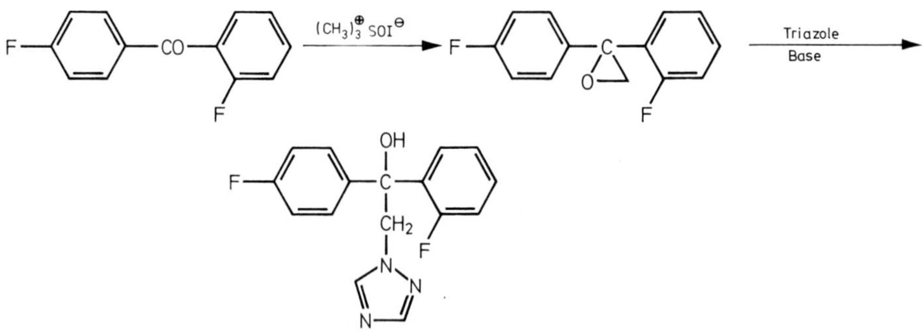

6.5 Azoles with Ether Groups

Etherification of the corresponding azolyl-carbinols also yields highly active products.

Imazalil® has been introduced as a seed treatment and fungicide.

1-[2-(2,4-Dichlorophenyl)-2-(2-propenyloxyethyl)]-1H-imidazole [107]

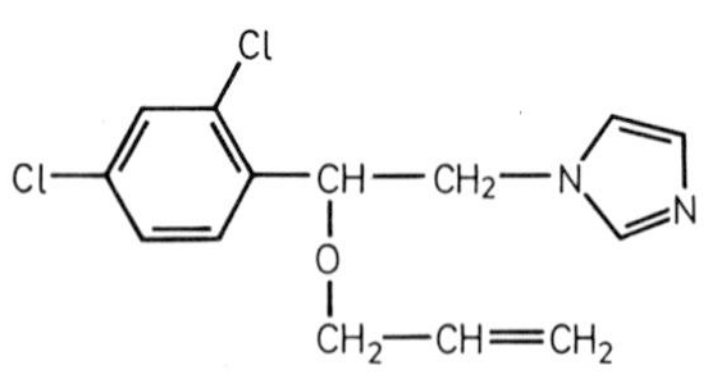

Imazalil®
Janssen
(and other companies as a partner in combination fungicides)
$LD_{50_{rat, oral, acute}}$: 343 mg/kg (male)
227 mg/kg (female)

This compound, an oily liquid with a water-solubility of 293 mg/l at 20 °C and high soluble in ethanol, methanol and xylene, is prepared from 2,4-dichloroacetophenone by halogenation, reduction, epoxide ring cleavage and etherification.

Alternatively the reduction can be carried out at the imidazolyl-ketone stage [108]:

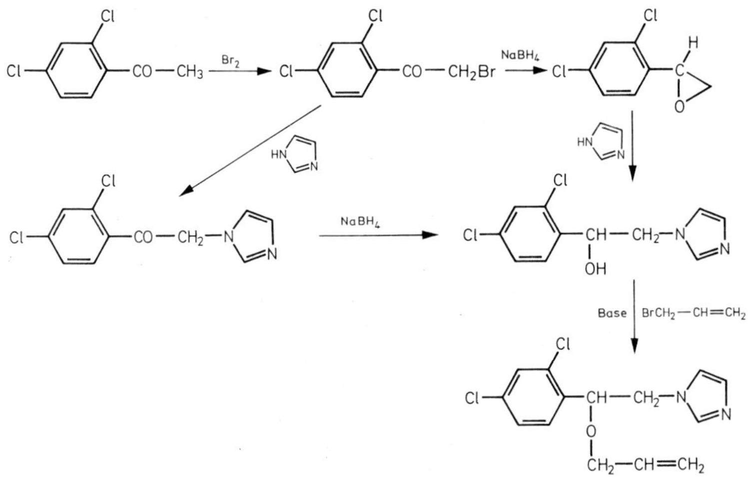

The antimycotics from the class 'Azoles with Ether Groups' are of great significance, both compounds with an ether and those with a thioether function. Miconazole, a topical antimycotic from Janssen Pharmaceuticals, was the forerunner of several other topical antimycotic which have already been introduced into clinical practice or are still undergoing trials.

The following table gives a survey.

Table 4. Topical Antimycotics from the Group 'Azoles with Ether-Functions

Structure	Name	Manufacturer	Literature
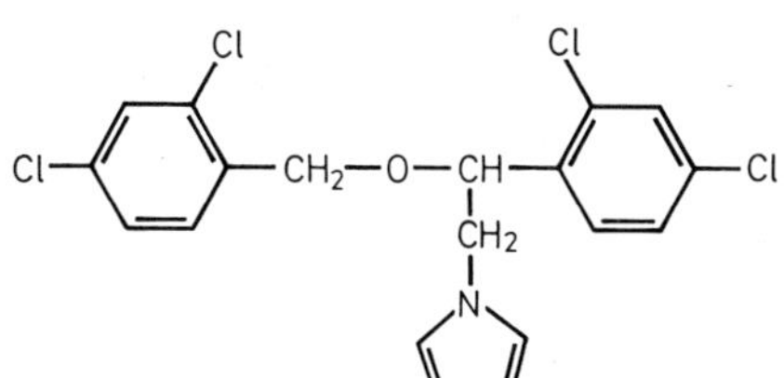	Miconazole Daktar® Monistat®	Janssen	DOS 1940388 (1967)
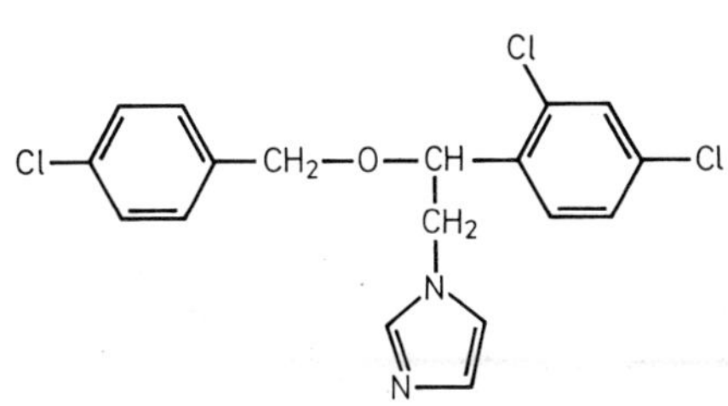	Econazole Pevaryl®	Janssen	DOS 1940388 (1967)

Table 4. (Continued)

Structure	Name	Manufacturer	Literature
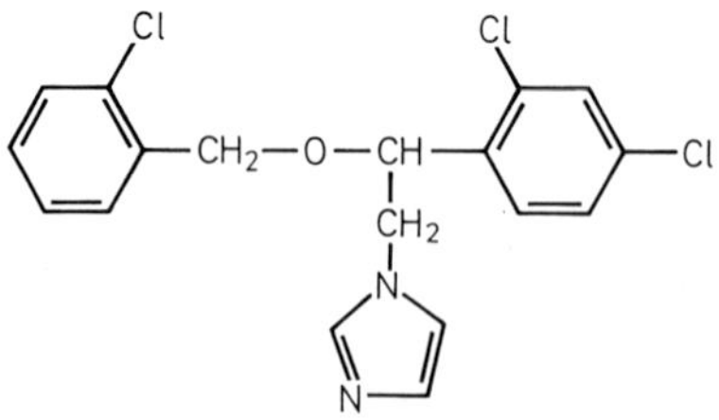	Isoconazole Travogen®	Janssen (Schering)	DOS 1940388 (1967)
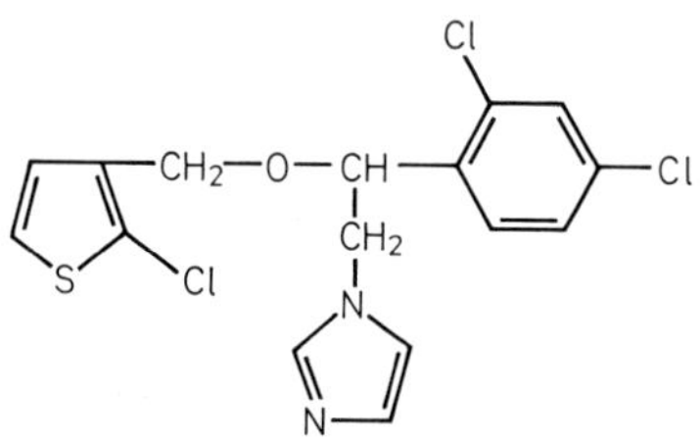	Tioconazole Fungipront® Proxyl® Trosyol®	Pfizer	DOS 2619381 (1976)
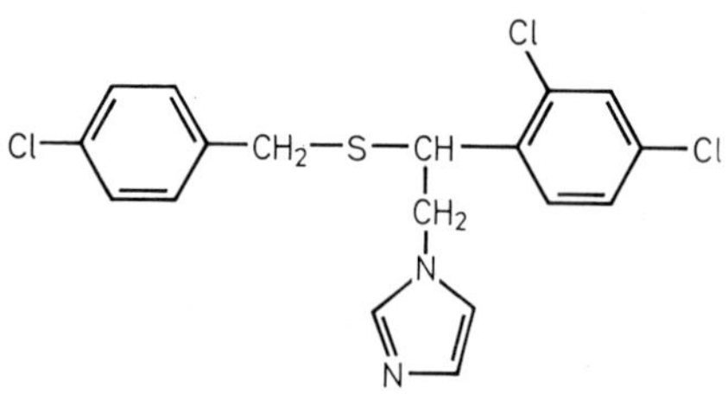	Sulconazole Sulcosyn®	Syntex	DOS 2541833
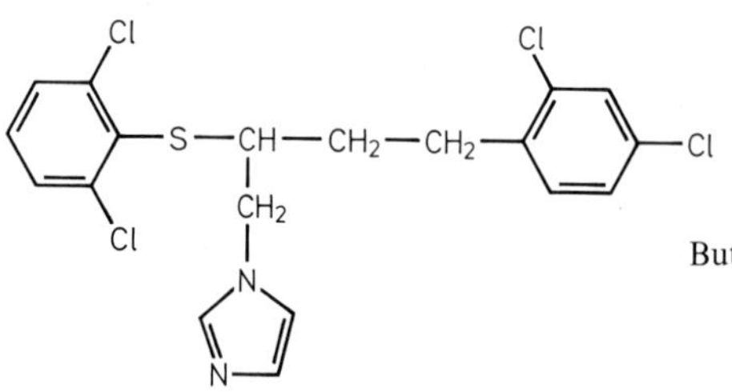	Butoconazole	Syntex	DOS 2633492 (1976)
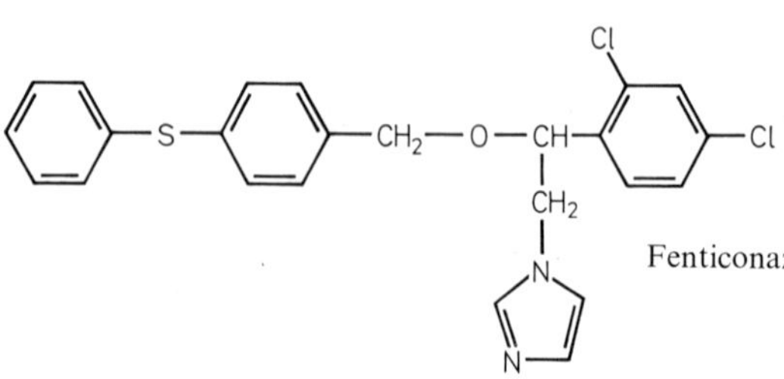	Fenticonazole	Recordati	DOS 2917244 (1978)

6.6 Azoles with Nitrilo Groups

Only one experimental product has been reported from this class to date:
α-Butyl-α-phenyl-1H-imidazole-1-propan-nitrile

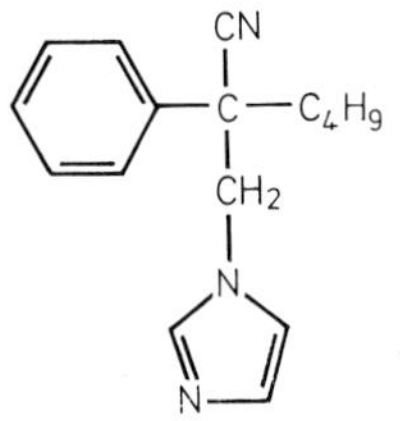

Fenapanil
Rohm and Haas
$LD_{50_{rat, oral, acute}}$: 1590 mg/kg

This compound, a colourless oil, is prepared from 2-cyano-2-phenyl-hexanol methanesulphonate and imidazole in DMF according to the following reaction scheme [108]:

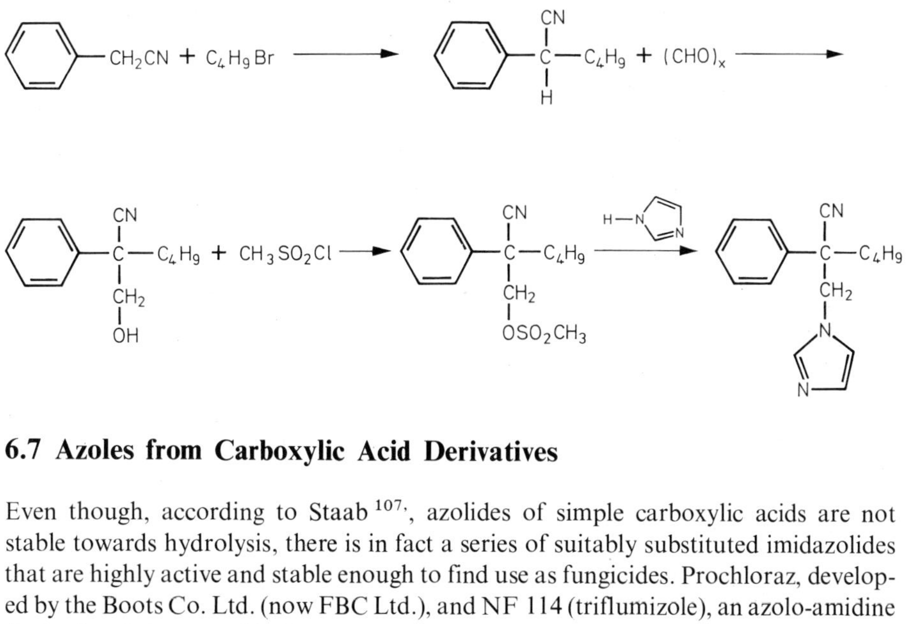

6.7 Azoles from Carboxylic Acid Derivatives

Even though, according to Staab [107], azolides of simple carboxylic acids are not stable towards hydrolysis, there is in fact a series of suitably substituted imidazolides that are highly active and stable enough to find use as fungicides. Prochloraz, developed by the Boots Co. Ltd. (now FBC Ltd.), and NF 114 (triflumizole), an azolo-amidine derivative, belong to this class.

N-Propyl-N-[2-(2,4,6-trichlorophenoxy)-ethyl]-1H-imidazole-1-carboxamide [110]

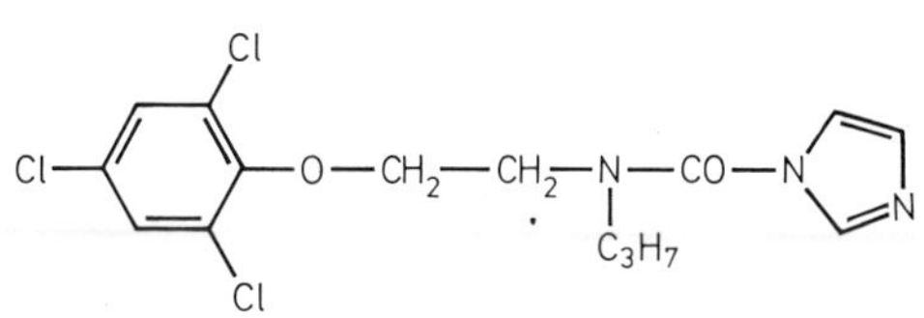

Prochloraz
Sportak®
Boots Co. Ltd.
$LD_{50_{rat, oral, acute}}$: 1600 mg/kg

This compound, with a melting point of 38.5–41 °C, is very slightly soluble in water (48 mg/l), freely soluble in organic solvents [111], and prepared by a normal urea synthesis from reaction of N-propyl-N-[2-(2,4,6-trichlorophenoxy)-ethyl]-carbamoyl chloride with imidazole [110].

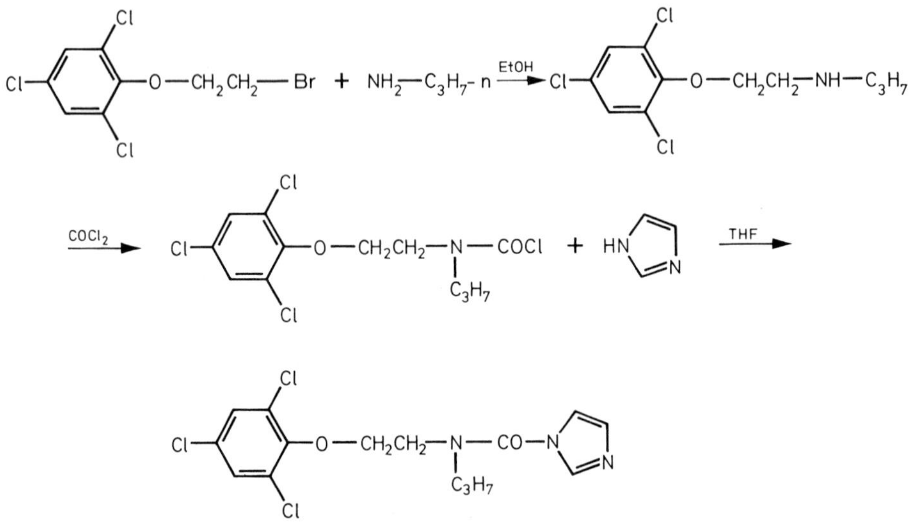

4-Chloro-N-[1-(1H-imidazol-1-yl)-2-propoxyethylidene]-2-trifluoromethyl-benzeneamine [112]

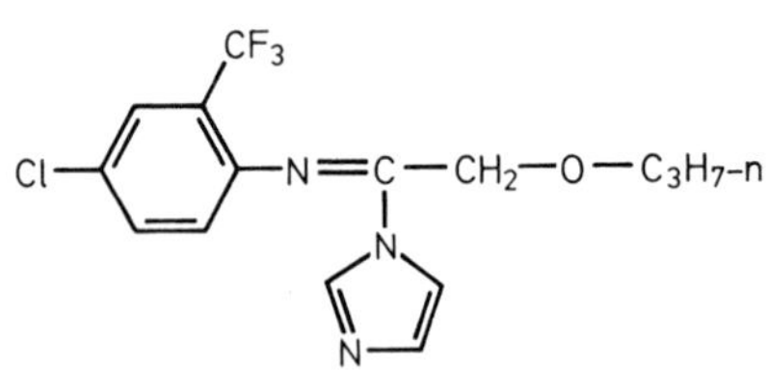

NF 114
Triflumizole (proposed)
Nippon Soda

a widely active experimental product from Nippon Soda, is obtained in an analogous fashion via the imide chloride by reaction with imidazole.

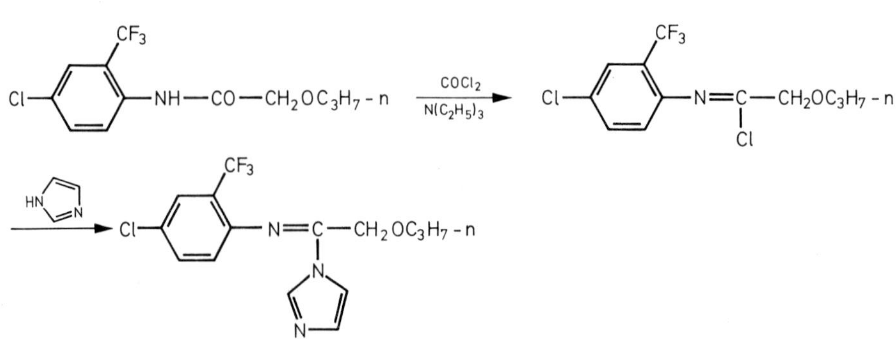

7 Amino Derivatives

In contrast to the 14-C-demethylase inhibitors of the pyridine, pyrimidine and azole type, the fungicidal and antimycotic morpholines, piperidines and naphthalene-methanamines interfere with ergosterol biosynthesis at various other intermediate steps [48, 115, 115a]. This substance group embraces both the oldest sterol biosynthesis inhibitors on the market, such as tridemorph, and the most recent experimental products, the antimycotics naftifine and SF 86-327.

7.1 Morpholines and Piperidines

2,6-Dimethylmorpholine derivatives substituted by acyclic or cyclic hydrocarbon residues of more than 12 carbon atoms have been known as fungicides for a long time, e.g. dodemorph, aldimorph, tridemorph [113].

More recent developments at BASF and Hoffmann-LaRoche have led to 3-phenyl-propylamines of the fenpropimorph and fenpropidine [115, 116] type and to the antimycotic R0-14-4767/002 under development at Hoffmann-LaRoche.

N-Cyclododecyl-2,6-dimethylmorpholine acetate [117]

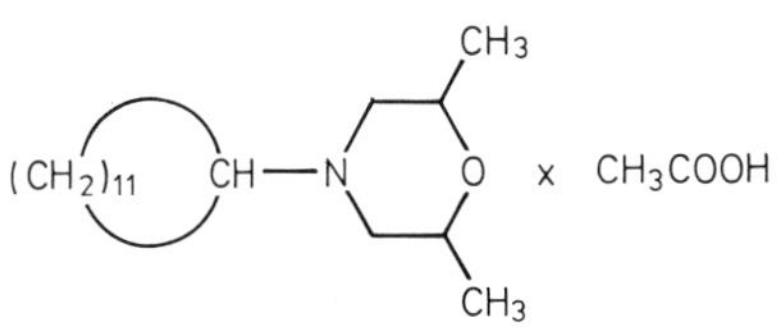

Dodemorph
Dodemorph®
BASF AG
$LD_{50_{rat, oral, acute}}$: 4800 mg/kg

Dodemorph, which occurs as a mixture of axial-equatorial and diequatorial forms, has a refractive index n_D^{25} of 1.4907 as the free base. It is only used in ornamental plants. The acetate has a low water solubility but is soluble in organic solvents.

α-Tridecyl-2,6-dimethylmorpholine

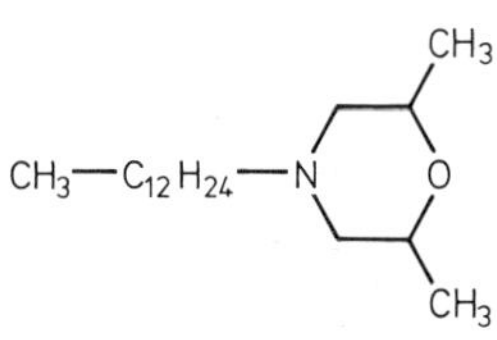

Tridemorph
Calixin®
BASF AG
$LD_{50_{rat, oral, acute}}$: 1000 mm³/kg (techn. ingr.)

has been marketed since 1965 as a cereal spray for control of mildew. It has a boiling point of 134 °C at 0.7 mbar, a water solubility of 0.01 g/100 g at 20 °C and is miscible in all proportions with ethanol, acetone, ethyl acetate, cyclohexane and chloroform. The last stage of the synthesis involves a dehydrative cyclisation of tridecyl-bis-(2-hydroxypropyl)-amine.

4-n-Dodecyl-2,6-dimethylmorpholine

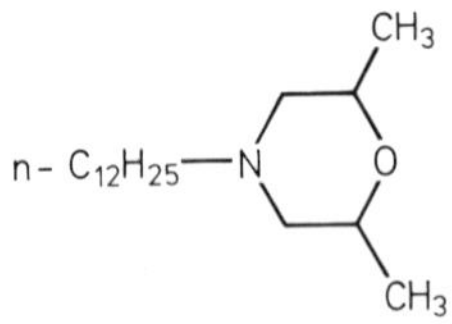

Aldimorph
Falimorph®
VEB Fahlberg List.
$LD_{50_{rat, oral, acute}}$: 3500 mg/kg

Aldimorph (proposed) is a mixture of various N-n-alkyl-dimethylmorpholines in which the main component is the dodecyl derivative. It has been introduced in East Germany as a cereal spray.

Compared with the classical 2,6-dimethylmorpholines an increase in the level of activity has been achieved with the 3-phenylpropylamines, partly by giving careful consideration to the stereochemistry of the 2,6-dimethylmorpholine moiety. The 3-phenylpropylamines of the fenpropimorph type contain cis-2,6-dimethylmorpholine, a much more active configuration than the corresponding trans form.

4-[3-[4-(1,1-Dimethylethyl)-phenyl]-2-methyl]-propyl-2,6-(cis)-dimethylmorpholine [119]

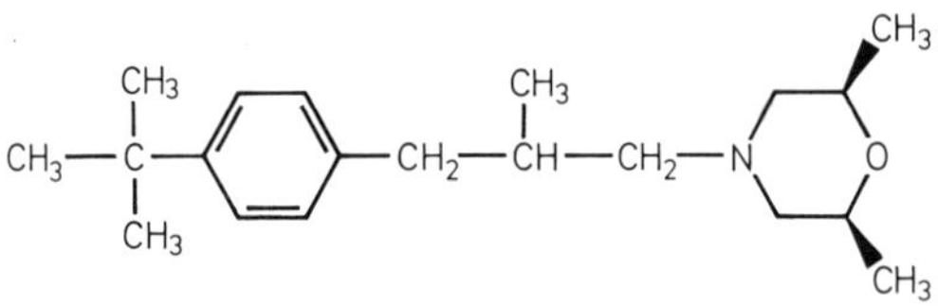

Fenpropimorph
Corbel®
BASF AG/Dr. R. Maag AG
$LD_{50_{rat, oral, acute}}$: 3515 mg/kg

Fenpropimorph has a boiling point of 120 °C at 5×10^{-2} mbar, a vapour pressure of 3.2×10^{-5} mbar at 20 °C and a water solubility of 6.8 mg/l at 20 °C. It is completely miscible with organic solvents. Starting materials for the synthesis are 3-(4-tert.-butyl-phenyl)-2-methylpropanal and cis-2,6-dimethylmorpholine. Fenpropimorph is obtained by reductive amination according to Leuchardt-Wallach.

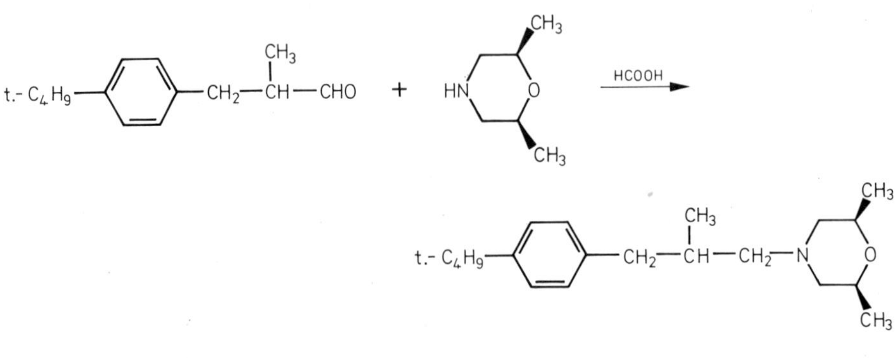

Structure-activity investigations in the phenylpropylamine class have shown biological activity to depend upon chain length of the alkyl bridge between the aromatic ring and the amine, the side-chain substitution of this alkyl bridge, and the substitution in the aromatic ring and the amine part [114].

High activity is also maintained upon replacement of the cis-2,6-dimethylmorpholine moiety by piperidine residues (e.g. fenpropidine), and particularly by perhydrogenation of the aromatic ring to give the trans-cyclohexyl compound:

N-[3-(trans-4-tert.-butylcyclohexyl)-2-methyl)-propyl]-piperidine [114)]

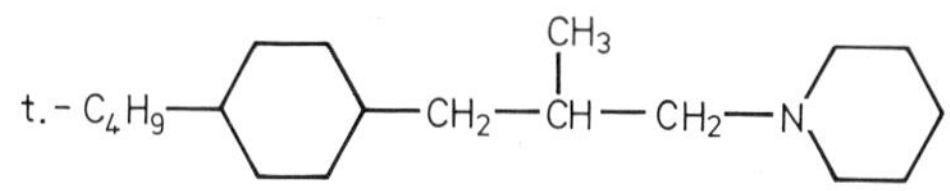

BASF AG

is an experimental product from this class [114)].

4-[3-[4-(1,1-Dimethylpropyl)-phenyl]-2-methyl]-propyl-2,6-(cis)-dimethylmorpholine [120)]

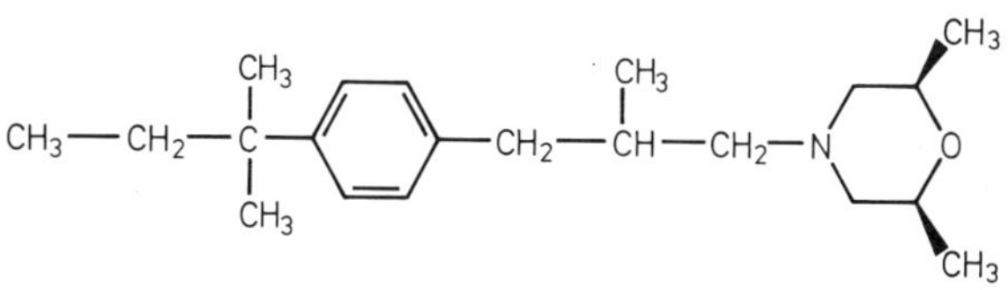

R0-14-4767/002
Hoffmann LaRoche

is currently undergoing clinical trials as a topical antimycotic. Replacement of the 4-butyl groups in fenpropimorph by the trimethylsilyl group also yields highly active products [121)].

7.2 Naphthalene-methanamines

Two experimental antimycotics are already known from this new group of sterol biosynthesis inhibitors that block squalene epoxidase [122)]:

(E)-N-Methyl-N-(3-phenyl-2-propenyl)-1-naphthalenemethanamine-hydrochloride [123)]

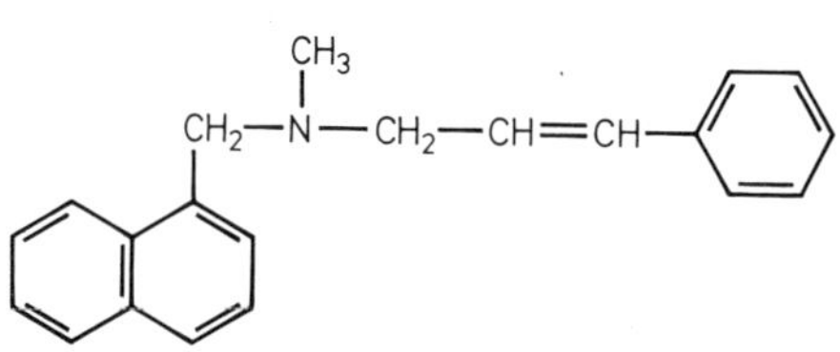

Naftifine
Sandoz AG

and

(E)-N-(6,6-dimethyl-2-hepten-4-ynyl)-N-methyl-1-naphthalenemethanamine-hydrochloride [124)]

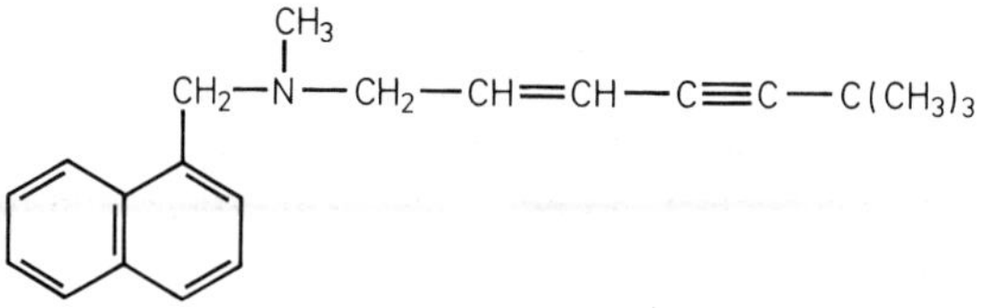

SF 86-327
Sandoz AG

Table 5. Stereoisomerism and Biological Activity of Azole Fugicides

Compound	Number/kind of isomeres	Determination data	Antifungal activity
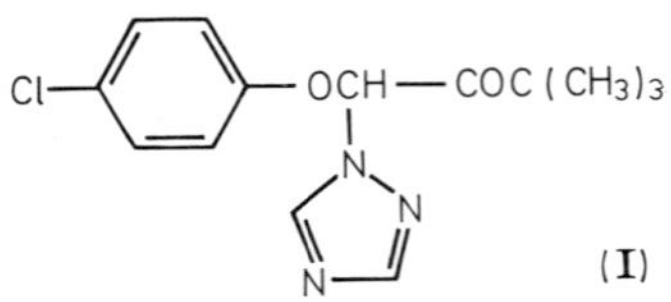 (I)	2 optical enantiomers	$[\alpha]_D^{23} = (\pm)54{,}4°$ CCl_4	+ ~ —
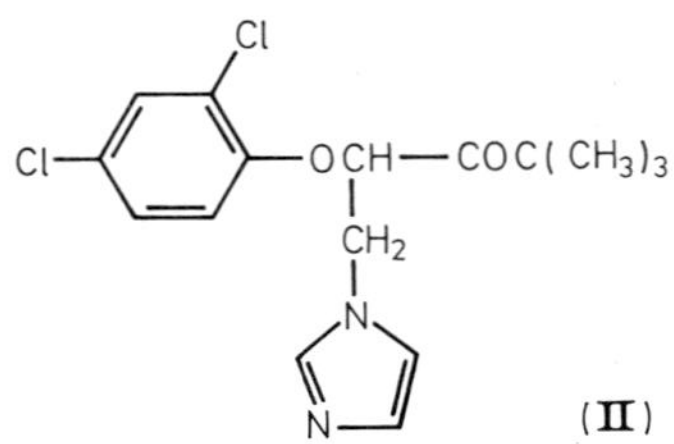 (II)	2 optical enantiomers	$[\alpha]_D^{20} = (\pm)39{,}4°$ (H_2O/C_2H_5) (95:5)	+ > —
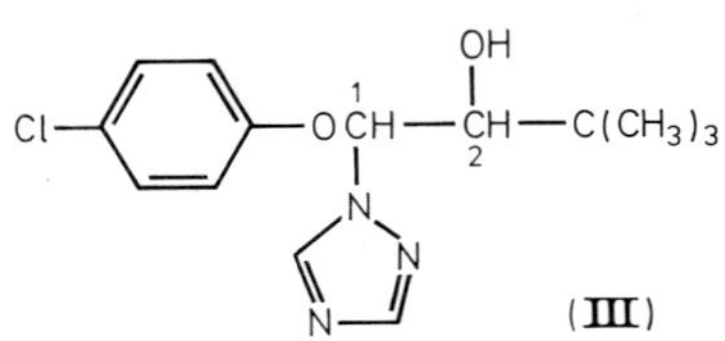 (III)	2 diastereomeric pairs 4 enantiomers	RS; SR $[\alpha]_D^{23} = \pm 124{,}4°$ $(CHCl_3)$ SS; RR $[\alpha]_D^{23} = \pm 140{,}6°$ $(CHCl_3)$	(—)1S,2R > (+)1R,2R (—)1S,2S > (+)1R,2S

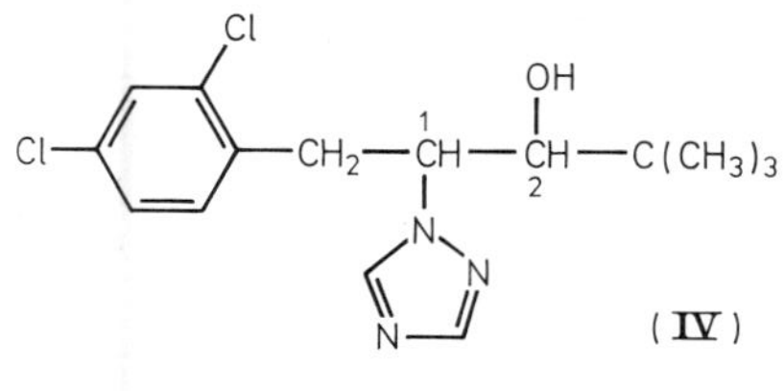 (IV)	2 diastereomeric pairs 4 enantiomers		1R,2R + 1S,2S > 1S,2R + 1R,2S 1R,2R > 1S,2S
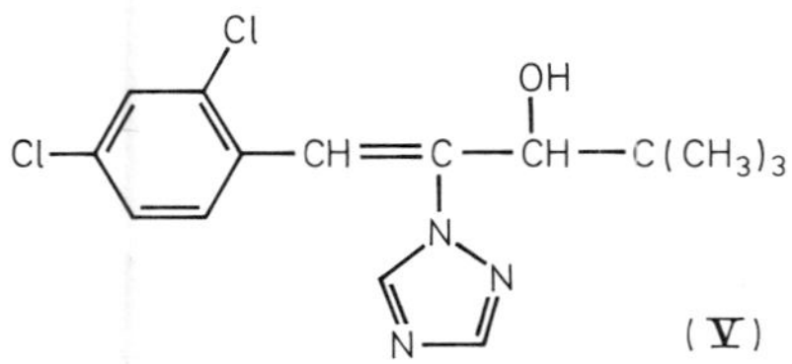 (V)	2 geometrical 4 enantiomers	E $[\alpha]_D^{24} = 31{,}7°$ ($CHCl_3$) Z	R_1 S_1 (+)E > (—)E E > Z
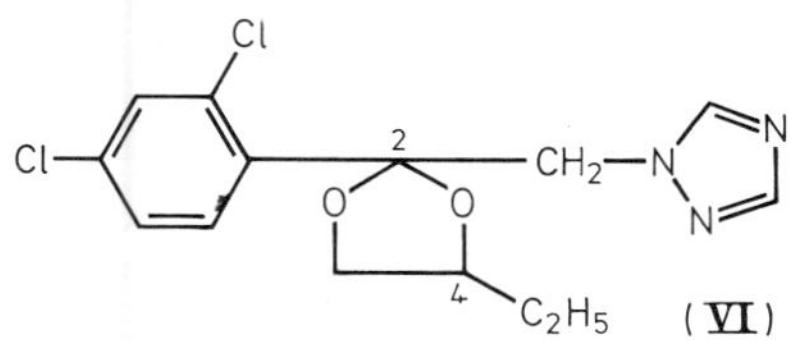 (VI)	2 geometrical 4 enantiomers	2S,4R; 2R,4S $[\alpha]_{545}^{20} = \pm 16 \pm 1°$ 2S,4S; 2R,4R $[\alpha]_{545}^{20} = -4 - 1°$	(—)2S,4R > (+)2R,4S (+)2S,4S > (—)2R,4R

Naftifine is an antimycotic for the topical treatment of dermatophyte infections [125]. SF 86-327 is undergoing clinical trials as an oral antimycotic for the treatment of systemic mycoses [126, 127]. Structure-activity studies in this class have indicated that, apart from the stereochemistry of the double bond (the trans configuration is essential for high activity), the chain length is of crucial importance for activity. The cinnamyl group of naftifine can be replaced by alkadienyl side-chains of corresponding chain lengths and by alkenine groups bearing branched alkyl groups at the acetylene function [128].

8 Stereochemical Implications

As has been described in the previous sections, the stereochemistry of the sterol biosynthesis inhibitors, particularly of the azoles and amines, can have a decisive influence upon biological activity.

Practically all the azoles marketed to date or in the experimental stage have at least one chiral carbon atom. The question concerning the influence of stereochemistry upon biological activity is therefore of particular interest in this class. Results in this area have been reported by several research groups.

In the case of triadimefon (I) it is surprising that no significant differences in activity were found considering that the MIC values of the closely related 1-imidazolyl-2-(2,4-dichlorophenoxy)-4,4-dimethylpentan-3-one enantiomers (II) differ by a factor of 4 [129, 130]. With triadimenol (III) the biological activity of the threo diastereomeric pair is higher than that of the erythro diastereomeric pair. The most active enantiomer is the threo (—)-1S,2R enantiomer, followed by the erythro (+)-1R,2R enantiomer [129].

In the case of dichlobutrazole (IV) the diastereomeric pair 1R,2R and 1S,2S is much more active than the 1S,2R and 1R,2S pair both in an S_8 yeast enzyme system (an enzymatic assay designed to measure the relative activities of inhibitors of sterol 14-demethylation) and in terms of their ED_{50} values against Ustilago maydis (0.2 μM 83 μM). In the S_8 yeast enzyme system the 1R,2R enantiomer is the most active of all; this correlates with the biological findings [131, 131a].

The N-vinyl-azole structure of the S 3308 type (V) encompasses two geometrical stereoisomers, E and Z, each of which has two enantiomers. The E isomers are more active than the Z. Surprisingly, the antifungal activity of the (—) E isomer is the greatest but the (+) E isomer has the greatest plant growth regulating activity [132, 132a].

The etaconazole (VI) structure also has 4 enantiomers with some differences in activity among them. In general the 2S,4R isomer is the most active one but the ranking of the isomers also varies depending on the test fungus [133].

The two enantiomers of fenpropimorph, separated by the classical technique with camphorsulphonic acid, also have different biological activities. The S(—)-enantiomer is much more active than the R(+)-enantiomer against mildew and orange leaf rust of wheat [114].

Research on sterol biosynthesis inhibitors thus does not just cover their chemical synthesis and structural optimisation but also the separation of diastereomers and enantiomers, and investigations on stereoselective syntheses.

9 References

1. Ragsdale, N. N. in Antifungal Compounds, Vol. 2, 333; Ed.: Siegel, M. R. Sisler, H.-D. Marcel, Dekker Inc. New York, 1977
2. Ragsdale, N. N., Sisler, H. D. Pestic Biochem. Physiol. *3*, 20 (1973)
3. Buchenauer, H. Z. Pfl. Krankh. Pfl. Schutz *83*, 363 (1976)
4. Kerkenaar, A., Barug, D., Kaars Sijpestein, A. Pestic Biochem. Physiol. *12*, 195 (1979)
5. Sherald, J. L., Ragsdale, N. N., and Sisler, H. D. Pestic. Sci. *4*, 719 (1973)
6. Sherald, J. L., Sisler, H. D. Pestic Biochem. Physiol. *5*, 477 (1975)
7. Kato, T., Tanaka, S., Ueda M., and Kawase, Y. Agr. Biol. Chem. (Tokyo) *31*, 169 (1975)
8. Ryder, N. S. SF-86-327: Studies on the antifungal mode of action, 13th International Congress of Chemotherapy, Vienna, 28. 8.–2. 9. 1983, Abstracts 4.8/4–6
9. Ziogas, B. N., Sisler, H. D., Lusby W. R. Pestic. Biochem. Physiol. *20*, 320 (1983)
10. Zirngibl, L. Progress of Drug Research, Vol. 27, 253–383 (1983), Birkhäuser Verlag, Basel, Boston
11. DOS 19 01 421 (1968) [CH. Boehringer und Sohn, Ingelheim] = NE 69 005 95 Inv.: Ost N., Thomas, K., Jerchel, D., and Appel, K.
12. Schicke, P., Veen, K.-H. Proc. Br. Insectic, Fungic. Conf., 1969, *5*, 569
13. Tanaka, S., Kato, T., Yamamoto, S., Yoshioka, H. Agric. Biol. Chem. (Tokyo) *41*, 1953 (1977)
14. DOS 21 19 174 (1970), Sumitomo Chem. Co. Ltd. Ind. Inv. Tanaka, S., et al., C.A. *76*, 46101f (1972)
15. JP 72/43334 (1971), Sumitomo Chem. Co. Ltd. Inv. Tanaka, S., et al., C.A. *78*, 144, 283c (1973)
16. Tanaka, S., Kato, T., Yamamoto, H., Yoshioka, H. Agric. Biol. Chem. *41*, 1627 (1977)
17. JAP 75 46 832 (1974), Sumitomo Chem. Co. Ltd. Inv.: Tanaka, S. et al., C.A. *83*, 92365n (1975)
18. Brit 1 175 693 (1966), Lilly, Eli and Co. Inv.: van Heyningen, E. M., Taylor, H. H., C.A. *72*, 78897d (1970)
19. Wright, W. L. Chem. Spec. Mfr. Ass. Proc. Annu. Meet. *1968*, (55) 147; C.A. *71*, 69583a (1969)
20. Whaley, J. W., Taylor, H. M. Phytopathology *1970*, 60 (5), 771
21. Kress, T. J., Moore, L. L. J. Heterocyclic Chem. *9*, 1161 (1972)

22a. EP 62 238 (1982), Hoffmann La Roche Inv.: Dorn, F

22b. EP 74 018 (1981) Hoffmann La Roche Inv.: Dorn, F., Montavon, F., Sachy, M.

23. IL 53 744 (1980), Eli Lilly and Co. C.A. *95*, 163890r (1981)
24. BR 78 00 036 (1979), Eli Lilly and Co. C.A. *93*, 199233b (1980)
25. NL 78 00 052 (1979) Eli Lilly and Co. C.A. *92*, 53384j (1980)
26. US 4 116 665 (1978) Eli Lilly and Co., Inv.: Krumkalns, E. V.
27. Sisler, H. D., Ragsdale, N. N., Waterfield, W. F., Pestic. Sci. *15*, 167–176 (1984)

27a. Coolbough, R. C., Swanson, D. J., West, Ch. A. Plant Physiol. *1982*, (69) 707

28. Fr. 1,569 940 (US-Priorität 1967), Eli Lilly and Co. Inv.: Davenport, J. D., Hackler, R. E., Taylor, H. M. C.A. *72*, 100745b
29. Martin, H. Pesticide Manual British Crop Protection Council, 4. Edn. London 1974
30. Hardison, J. R. Plant Disease Reptr. *56*, 55 (1972)
31. Pesticides Process Encyclopediap. 20
32. Coolbough, R. C. J., Heil, D. R., West, Ch. A. Plant Physiol. *1982* (69), 712
33. Barak, E., Dinoor, A., Jacoby, B. Proc. Br. Crop. Prot. Conf. Pests, Dis. *1981*, 11th, 129
34. Staden, O. L., Witmondt, M., Proc. Strawberry Growing 1970'S, 74, Baywood Chem. Ltd., Bury St. Edmunds 1969
35. Pawlak, J. et al. Tetrahedron Lett. (1979), 1533
36. Patterson, J., Holland, J., and Bieber, L. L. J. Antibiot. *32*, 646 (1979)
37. Montesanto, R., Nature (London), *280*, 328 (1979)
38. Bauer, K., et al., Chemie der Pflanzenschutz- und Schädlingsbekämpfungsmittel, Band 6, 272, Ed.: R. Wegler, Springer Verlag Berlin, Heidelberg 1980
39. Golding, B. T. et al. Tetrahedron Lett. (1966), 3551
40. Ceder, D., et al. Chimia *18*, 83 (1964)
41. Dekker, J., World Rev. Pest. Control *10*, 9 (1971)
42. Golding, B. T., et al. Tetrahedron Lett. *37*, 2615 (1964)
43. Otten, H., Plempel, M., Siegenthaler, N. Antibiotika Fibel, Georg Thieme Verlag Stuttgart, 4. Edn. p. 666

44. Antibiotika Fibel, Georg Thieme Verlag Stuttgart, 4. Edn. p.694 Ed.: Otten, H., Plempel, M., Siegenthaler, N.
45. McCammon, M. T., and Perks, L. W. J. Bacteriol. *145*, 106 (1981)
46. Hamill, R. L., Hoehn, M. M. US 3 758 681, 11. 9. 1973, Eli Lilly and Co.
47. Nagarajan, R. US 415 80 56, 12. 6. 1979; Eli Lilly and Co.
48. Baldwin, B. C., Biochem. Soc. Trans. *11*, 659 (1983)
49. Howard, W. R., Hoehn, M. M., Michel, K. H. BE 79 518, 29. 2. 1972; Eli Lilly and Co. = DOS 2 309 108 (2. 2. 1973)
50. Rodriguez, R. J., Park, C. W. Antimicr. Agents and Chemother. *20*, 184 (1981)
51. Arruch, L., Fisler, S., Pierce, Hr. jr., Oehlschläger, A. C. Can. J. Biochem. *54*, 657 (1976)
52. Fryberg, M., Oehlschläger, A. C., Unrau, A. M. J. Am. Chem. Soc. *95*, 5747 (1973)
53. Pierce, H. D. jr., Pierce, A. M., Srinivasan, R., Unrau, A. M., Oehlschläger, A. C. Biochim. Biophys. Acta, *529*, 429 (1978)
54. Fryberg, M., Oehlschläger, A. C., Unrau, A. M. Tetrahedron *27*, 1261 (1971)
55. Belg. Pat. 720 801 (1968), Bayer AG Inv.: Büchel, K. H., Regel, E. and Plempel, M.
56. Plempel, M., Bastmann, K., Büchel, K. H. and Regel, E. Deutsche Med. Wochenzeitschrift *94*, 1356 (1969)
57. Go defroi, E. F., Heeres, J., Van Cutsem, J., Janssen, P. A. J. J. Med. Chem. *12*, 784 (1969)
58. Büchel, K. H., Draber, W., Regel, E., Plempel, M. Arznm. Forschung (Drug Research) *22*, 1260 (1972)
59. DOS 1 670 976 (1968) Bayer AG Inv.: Büchel, K. H., Regel, E., Grewe, F., Scheinpflug, H., Kaspers, H.
60. DOS 2 063 857 (1971) Janssen Pharmaceutica Inv.: Godefroi, E. F., Schuermans, J. L.
61. DOS 1 940 626 (1969) Bayer AG Inv.: Jäger, G., Büchel, K. H.
62. DOS 2 009 020 (1971) Bayer AG Inv.: Draber, W., Regel, E.
63. DOS 1 940 628 (1971) Bayer AG Inv.: Büchel, K. H., Draber, W.
64. Büchel, K. H., Draber, W., Regel, E., Plempel, M. Drugs made in Germany *15*, 79—94 (1972)
65. Maier, F., Mildenberger, H., Angew. Chem. *92*, 128—129 (1980)
66. DOS 2 610 022 (1975) ICI
67. EP 8 804 (1979) Siegfried AG Inv.: Zirngibl, L., Fischer, J., Thiele, K.
68. DOS 2 931 755 (1981) Bayer AG Inv.: Draber, W., Büchel, K. H., Frohberger, P. E., Brandes, W.
69. DOS 2 549 798 (1975) Nordmark Inv.: Gutsche, K., Kohlmann, F. W., Scharwächte: P
70. DOS 2 652 313 (1976) BASF AG Inv.: Pommer, E. H., Gutsche, K., Hartleben, Y.
71. Jap. Pat. J 53 130 661 (1977) Sumitomo Chem. Co. Inv.: Funaka, Y., et al.
72. DOS 3 032 326 (1980) Bayer AG Inv.: Krämer, W., Büchel, K. H., Brandes, W., Frohberger, P. E.
73. Krämer, W., Büchel, K. H., Meiser, W., Brandes, W., Kaspers, H., Scheinpflug, H. Advances in Pesticide Science, Edited by: Geissbühler, H.; Part 2, p274, Pergamon Press Oxford and New York, 1979
74. Gadher, P., Mercer, E. I., Baldwin, B. C., Wiggins, T. E. Pesticide Biochemistry and Physiology *19*, 1 (1983)
75. DOS 2 735 872 (1976) Janssen Pharmaceutica N. V. Inv.: Heeres, J., Backx, L. J., Mostmans, J. A.
76 FR 2 016 526 (1968) Bayer AG Inv.: Büchel, K. H.
77. DOS 2 546 532 (1975) Bayer AG Inv.: Klauke, E.
78. DOS 2 549 095 (1975) Bayer AG Inv.: Büttner, G., Klauke, E., Schwarz, H., Döhring, F.
79. EP 63099 (1981) Ciba Geigy AG Inv.: Maier, L., Kunz, W.
80. DOS 2 461 406 (1976) Bayer AG Inv.: Regel, E., Büchel, K. H., Draber, W., Metzger, C., Plempel, M.
81. DOS 2 044 621 (1972) Bayer AG Inv.: Jäger, G., Büchel, K. H., Plempel, M.
82. Berg, D., Regel, E., Harenberg, H. E., Plempel, M. Arzneimittel-Forschung (Drug Research) *34*, [1], 2, 139 (1984)
83. DOS 2 201 063 (1972) Bayer AG Inv.: Meiser, W., Büchel, K. H., Krämer, W., Grewe, F.
84. DOS 2 406 665 (1974) Bayer AG Inv.: Stölzer, C., Krämer, W., Büchel, K. H., Meiser, W.
85. Berg, D., Büchel, K. H., Krämer, W., IUPAC Pesticide Chemistry, Human Welfare and the Environment, Ed.: J. Miyamoto et al., Pergamon Press 1983, p297, Vol. 1
86. DOS 2 739 352 (1977) BASF AG, DOS 3 011 258 (1980) BASF AG Inv.: Rentzea, C., Ammermann, E., Pommer, H.

87. Rademacher, N., Jung, J., Z. Acker- und Pflanzenbau 150 (1981), 363—371
88. DOS 2 819 879 (1977) ICI Inv.: Shepard, M. C., Worthington, P. A., Maidenhead, B. J. J., Boice, L. M.
89. Shepard, M. C., French, P. N., Rathmell, W. G., 10th International Congress of Plant Protection 1983, Proceedings of the Congress, Brighton, November 20.–25., 1983, Vol. 2., p521
90. DOS 2 551 560 (1974) Janssen AG Inv.: von Reet, G., Heeres, J., Wals, L.
91. Speich, J., Urech, P. A., Meded. Fac. Landbouwwet. Rijksuniv. Gent 1980, Vol. 45 (2), 315
92. Paulus, A. O., Nelson, J., Besemer, S., Brendler, B., Hall, B., Meded Fac. Landbouwwet., Rijiksuniv. Gent. 1980, Vol. 45 (2), 245
93. Heeres, J., Backx, L. J. J., Mostmans, H., van Cutsem, J. Journal of Med. Chem. (1979), *22*, (8), 1003
94. Drugs Future *4*, 513 (1979)
95. (Janssen): J. Am. Med. Assoc. *237*, 2422 (1977)
96. (Janssen): J. Am. Med. Assoc. *243*, 367 (1980)
97. Borgers, M., van de Ven, M., Willemsens, G., van den Bosche, H., 13th ICC, Wien 1983, Abstracts SE 4.8/2–13
98. Pommer, E., Zeeh, B., 4. International Congress Plant Pathology, Melbourne, Australia (1983) Abstract
99. EP 5 794 (1978) Siegfried AG Inv.: Mixich, G., Thiele, K., Fischer, J.
100. Frohberger, P. E., Pflanzenschutz-Nachrichten Bayer *31*, 11—24 (1978)
101. DOS 2 324 010 (1973) Bayer AG Inv.: Krämer, W., Büchel, K. H., Meiser, W., Kaspers, H., Frohberger, P. E.
102. Brandes, W., Kaspers, H., Krämer, W., Pflanzenschutz-Nachrichten Bayer *32*, 1–16 (1979)
103. DOS 2 737 489 (1976) ICI Inv.: Balasubram, S., Shepard, S., Batch, J. J., Boize, L. M.
104. DOS 3 010 560 (1979) Sumitomo Chem. Co. Inv.: Funaka, Y., Oshita, H., Yamamoto, S., Tanaka, S., Kato, T.
105. Skidmore, A. M., French, P. N., Rathmell, W. G., 10th International Congress Plant Protection 1983, Brighton, Abstract Vol. 1-2C-11, p368–375
106. EP 15 756 (1979) ICI Inv.: Parry, K. P., Rathmell, W. G., Worthington, P. A.
107. BP 1 244 530 (1969) Janssen AG Inv.: Godefroi, E. F., Heeres, J.
108. DOS 2 604 047 (1976) Rohm and Haas Comp. Inv.: Miller, G. A., Carley, H. E., Chan, H. F.
109. H. Staab, Chem. Ber. *89*, 1927–1940 (1956)
110. US-Pat. 3 991 071 (1973) Boots Co. Ltd. Inv.: Brooks, R. F., Godson, D. H., Hams, A. F., Weight, D. M., Wells, W. H.
111. Birchmore, R. J., Brookes, R. F., Copping, G. L., Wells, W. H., Proceedings 1977, British Crop. Protection Conf. Vol. 2, p593
112. DOS 2 814 041 (1978) Nippon Soda Inv.: Ikura, K., Katsuura, K., Kataoka, M., Nakada, A., Mizumo, M.
113a. König, K. H., Pommer, E. H., Sanne, W., Angew. Chem. *77*, 327 (1965)
113b. Büchel, K. H., Chemistry of Pesticides, John Willey and Sons, New York, Chichester, Brisbane, Toronto, Singapore 1983, 303, 305
114. Himmele, W., Pommer, E. H., Angew. Chem. *92*, 176–181 (1980)
115. Kerkenaar, A., Pestic. Chem. Hum. Welfare Environ. Proc. Intern. Congr. Pestic. Chem. 5th 1982, (Pub. 1983), 3, 123–7. Ed.: Miyamoto, J., Kearny, P. C., Pergamon Press Oxford and New York
115a. Kerkenaar, A., et al. Pestic. Sci. 1984, *15*, 177–187
116. Leroux, P., Gredt, M., C. R. Seances Acad. Sci. Ser. 3 (1983), 296 (4), 191–4, C.A. 99, 100843 g
117. Belg. Pat. 61 42 14 (1961) BASF AG Inv.: Kradel, J., Pommer, E. H., Effland, H. C.A. 58, 4581f
118. Brit. Pat. 1 275 485 (1968) BASF AG, FR 2 022 193, US 3 649 627
119a. DOS 2 752 135 (1978) Hoffmann La Roche Inv.: Pfiffner, A., Bohnen, K.
119b. DOS 2 656 747 (1978) BASF AG Inv.: Himmele, W., Pommer, E. H., Goetz, N.
120. DOS 2 752 096 (1978) Hoffmann La Roche Inv.: Pfiffner, A., Bohnen, K.
121. EP 93 274 (1983) BASF AG Inv.: Acker, R. D., Buschmann, E., Pommer, E. H., Ammermann, E.
122. Paltauf, F., Daum, G., Zuder, G., Högenauer, G., Schulz, G., Seidl, G., Biochem. Biophys. Acta *712*, 268–273 (1982)
123. DOS 2 716 943 (1977) Sandoz AG Inv.: Berney, D.

124. Stütz, A., SF-86-327 and Related Compounds: Synthetic Methods, 13th Intern. Cong. of Chemotherapy, Vienna 1983, Abstracts PS 4.8/4–2
125. Ganzinger, U., Stephan, A., Gumhold, G., Clinical Trials J. *19*, 342–348 (1982)
126. Ganzinger, U., et al., SF 86–327, Evalution of Toxicity in Laboratory Animals, Tolerance and Pharmakokinetics after oral Application to Man. 13th International Congr. of Chemotherapy, Vienna, August 28th to September 2nd 1983, Abstracts PS 4.8/4–13
127. Petranyi, G., Meingassner, J. G., Mieth, H., SF 86–327, A new Antimycotic Agent and its Antifungal Activity in vitro. 13th International Congress of Chemotherapy, Vienna, 28. 8.–2. 9. 1983, Abstracts PS 4.8/4–3
128. Stütz, A., Petranyi, G., SF 86–327, and Related Allylamine Derivatives with enhanced oral Activity 13th International Congress of Chemotherapy, Vienna, 28. 8.–2. 9. 1983, Abstracts: PS 4.8/4–1
129. Krämer, W., Büchel, K. H., Draber, W., Pesticide Chemistry, Human Welfare and the Environement Vol. 1, 223—232 (1983)
130. DOS 2816818 (1979) Bayer AG Inv.: Schmidt, Th., Krämer, W., Büchel, K. H., Plempel, M.
131. Gadler, P., Mercer, E. I., Baldwin, B. C. Wiggins, T. E., Pestic. Biochem. and Physiol. *19*, 1–10 (1983)
131a. Baldwin, B. C., Wiggins, T. E., Pestic. Sci. *15*, 156–166 (1984)
132. EP 54431 (1980) Sumitomo Chem. Co. Inv.: Funaka, Y., et al.
132a. Funaki, Y., Ishiguiri, Y., Kato, T., Tanaka, S., Pestic. Chem. Human Welfare Environ. Proc. Intern. Congr. Pestic. Chem. 5th 1982 (Pub. 1983), Vol. 1, 309—314. Ed.: Miyamoto, Y., Kearney, P. C., Pergamon Press Oxford, UK.
133. Vogel, C., Schaub, T., Rist, G., Sturm, E., Pesticide Chemistry, Human Welfare and the Environment, Proc. Int. Congr. Pestic. Chem. 5th, 1983 (Pub. 1983), Vol. 1, 303–308. Ed.: Miyamoto, Y., Kearney, P. C., Pergamon Press Oxford, UK. C.A. 99,48850 g (1983)

Biology of Sterol-biosynthesis Inhibiting Fungicides

K. H. Kuck and H. Scheinpflug
Bayer AG, Geschäftsbereich Pflanzenschutz, Biologische Forschung, 4019 Monheim, FRG

The sterol biosynthesis inhibiting fungicides (SBI's) are among the most exciting recently developed agricultural chemicals. Their rapid increase in importance is based on a combination of properties, the most important of which are covered in detail in this article.

First, the broad spectrum of fungicidal activity observed in laboratory and practical studies will be discussed.In addition alterations in fungitoxicity caused by stereochemical differences will be examined.

Secondly, the uptake and translocation of these fungicides after spray or seed treatment, properties which allow their use in curative treatments, will be described.

The last aspect, the influence of SBI's on the physiology of plants has created considerable interest, not only among agriculturalists but also in the field of plant physiology. This is due to the fact that ordinary side effects such as incompatibility caused by the treatment, are often overshadowed by distinct plant growth regulator effects.

1 Introduction

Previous to the late sixties, the exclusive means of preventing fungal diseases were protective spray schedules with fungicides which generally lacked a pronounced curative or eradicative activity. Prominent examples for this type of "conventional" fungicide are copper- and sulfur-preparations, the dithiocarbamates, and the phthalimides.

Curative or eradicative control implies that a fungicide can combat a pathogen which has already successfully established itself in a host plant. The ability of a chemical to be effective in such cases implies two properties which are usually not found in combination in conventional fungicides.

First, the molecule must be able to enter the plant surface via shoot-, seed-, or root-surfaces; secondly, its mode of action must be highly specific for fungi and should not affect the host plant.

Whereas unspecific multisite-effectors have to be necessarly non-systemic to avoid plant damage, site-specific inhibitors can be systemic or non-systemic. As Fuchs et al. [1] pointed out, this statement of importance as site-specificity is often illegitimately used as a synonym for systemicity.

With the discovery of several new classes of systemic fungicides which fulfil the requirements for curative treatment, agriculture obtained powerful tools for the control of fungal diseases. The carboxamides, the MBC-generating fungicides, the hydroxypyrimidines and more recently the acylalanines are only a few examples of fungicide classes which have rapidly become of importance in the market for curative control.

The sterol-biosynthesis inhibiting fungicides (abbreviated SBI or SBI's), which belong to a confusing multiplicity of chemical classes, are by far the most rapidly growing group of systemic fungicides. This is particularly evident if one considers the number of new compounds which have been introduced in the last years, and their importance in the market. Although all the systemic fungicides taken together were responsible for only 8% of the world fungicide market in 1980, the SBI-fungicides alone covered 8% of this market in 1982; it is estimated that they will have a market share of 12–15% in 1985 [2].

The following reasons are mainly responsible for the success of this class of fungicides:

1. A manifold of leaf- and seedling diseases in commercially important crops can be controlled by SBI's. Important examples are the powdery mildews and the rust fungi; these obligate parasites cause problems in nearly every plant family. Due to the broad fungicidal spectrum of SBI's, not only a single pathogen, but in many cases a complex of different pathogens can be simultaneously controlled.
2. The systemic properties of most SBI's not only allow protective but also curative treatments of above ground parts of plants and principally allow their use as a systemic seed treatment.
3. Until now resistance problems have been of relatively low importance since resistance of fungi to SBI-fungicides seems to be correlated with reduced fitness of the pathogen [3].
4. The high fungicidal potential of SBI's allows the application of low doses.

This chapter will discuss the fungicidal spectrum, systemic properties and the influences on plant physiology of those SBI's which are in use or which presumably will become of importance. Although questions regarding the build up of resistance to SBI's have become of increasing interest during the last few years, the subject is outside the scope of this article since several detailed monographs dealing with this subject have been recently published [4,5,6].

2 Spectrum of Fungicidal Activity

2.1 General

At low to moderate concentrations, the germination of fungal spores is only partially inhibited by SBI-fungicides [7,8,9,10]. Thus, some fungal pathogens are able to infect treated plants. Several studies with SBI's belonging to different chemical groups have shown that a gradual inhibition at early stages of fungal developement often occurs. Thus, shortening of germtubes [7,10] and partial inhibition of appressoria formation [8,11] have been reported. In the case of the obligate parasites rust [11] and powdery mildew [12,13], the final and obviously most important event of fungicidal action is the inhibition of haustorial development. In the case of apple scab, the inhibition of mycelial growth is the important event [10,14].

It is of interest that after treatment of susceptible host plants with SBI's, histological events sometimes occur which microscopically bear a striking resemblance to the resistance reactions of untreated, but genetically resistant plants. Thus, Kuck et al. [11] found 'hypersensitive-like' reacting host cells in wheat leaves which were infected with stem rust (Puccinia graminis f. sp. tritici), whereas Smolka and Wolf [13] demonstrated an encapsulation of powdery mildew (Erysiphe graminis) haustoria in epidermal wheat cells after treatment of the plants with triadimefon. It is therefore conceivable that the action of SBI-fungicides, which — at least at low concentration — are by virtue of their mechanism of action fungistatics rather than fungicides, is sometimes restricted to weakening the pathogen to such an extent that the host plant can overcome it.

The obligate parasites powdery mildew and — to a lesser extent — rust are clearly the basic targets of SBI-fungicides. Beyond that, fungi of nearly all important systematic categories are more or less sensitive to SBI's. The only important exception to this are the Oomycetes: of these the pythiaceous fungi Pythium and Phytophthora, which include many important plant pathogens, use plant sterols as precursors for their own biosynthetic needs [15,16]. Because of this, inhibitors of sterol biosynthesis do not appear to have their usual target. Nevertheless, the Oomycetes may possess unknown sites of attack; with several of these fungi an in vitro-inhibition by SBI's has been reported [17,18].

2.2 Piperazines, Pyridines and Pyrimidines

The piperazine — derivative *triforine* was described for the first time in 1971 by Celamerck [19]. Well known trade names are Saprol® and Funginex®. The spectrum of fungicidal activity in vitro and in vivo was described in detail by Drandarewski [20] and Fuchs and Drandarewski [21].

The most important fields of application of triforine are powdery mildew diseases in fruits, vegetables and ornamentals. For example, the compound is effective at 0.014–0.048% a.i. against Podosphaera leucotricha on apple, Sphaerotheca pannosa on roses, Erysiphe cichoracearum on cucumber, Uncinula necator on grapevine and Oidium mangiferae on mango [22)].

In cereals, triforine is used at 190–285 g a.i./ha against powdery mildew in barley and wheat. In addition to the EC- and WP-formulations, which are usually applied for foliar treatments, a liquid seed treatment formulation was developed against powdery mildew in spring barley in Great Britain [23)].

Among the rust fungi, Puccinia-, Uromyces- and Phragmidium-species in cereals, fruits and ornamentals are controlled by triforine. According to Drandarewski and Schicke [7)], triforine also has curative properties against apple scab (Venturia inaequalis).

There are many reports on efficacy of triforine against Monilinia diseases (brown rot, blossom blight) in peaches, apricots, cherries and other fruit trees [24, 25, 26)]. On blueberry, triforine is effective against mummy berry disease (Monilinia vaccinia-corymbosi) [27, 28, 29)].

Finally, some leaf spot diseases caused by Colletotrichum-, Cladosporium- and Isariopsis-species on beans, tomatoes and other vegetables are also controlled by triforine [30, 31, 32)].

Unlike most other SBI-fungicides, triforine has a clear side-effect on spider-mites (Panonychus ulmi) on apple [33, 34)]. Since the triforine molecule has two chiral C-carbons, the technically active ingredient is a mixture of three stereo-isomers: about 50% consits of meso-triforine whereas dextro- and laevorotary triforine make up the rest (about 25% each) [35)]. According to Ost [35)], the fungicidal activity declines in the order levorotary triforine ≫ meso-trif. > dextroratary triforine.

Buthiobate and its fungitoxic properties were first described by Kato et al. [36, 37)]. Compared to other SBI-fungicides, buthiobate has a narrow usable spectrum of fungitoxicity. Although Kato et al. [37)] reported low ED 50-values in vitro against some fungi within the Ascomycetes (f.e. Monilinia spp., Cochliobolus spp.) and some Fungi Imperfecti, the practical application of buthiobate is restricted to powdery mildew. Preventive and curative foliar sprays at concentrations from 0.007 to 0.02% a.i. are used to control Sphaerotheca fuliginea on cucurbits, Erysiphe cichoracearum on egg plants, Podosphaera leucotricha on apple and several other powdery mildews in ornamentals, vegetables and fruits [38)].

Triarimol was the first of a group of three systemic pyrimidine-fungicides presented by Eli Lilly and Co. which basically have similiar fungicidal properties. The compound was in official tests in 1969, but for toxicological reasons was suspended from further development in 1972 [39)]. Triarimol was an effective fungicide in apple mildew. In addition, it gave an adequate control of apple scab. Other crops on which triarimol controlled powdery mildews included roses and other ornamentals, cucumbers, grapes, strawberry, blackcurrent and gooseberry.

The second compound of this group is *fenarimol*. The usuable spectrum of this fungicide is similar to that of triarimol. Comparative in vitro studies with other SBI-fungicides such as triadimefon, triadimenol, imazalil and nuarimol, published by Buchenauer [40)], revealed a broad spectrum of activity against fungi of various systematic categories. As expected, there was a strong similarity to the structurally related compound nuarimol. In contrast to nuarimol, which is preferred for use in cereal diseases, fenarimol was developed mainly in fruits and vegetables.

In apples, fenarimol, the trade names of which are Rubigan® or Rimidin®, is effective in protective and curative treatments against powdery mildew and scab [41, 42, 43], in grapes it is effective against Uncinula necator [44]. Other powdery mildews which can be controlled by fenarimol are those of roses, tomatoes, peppers, cucurbits and peaches [41]. The concentrations necessary for control are generally low, ranging from 0.0018 to 0.0036% a.i. In some regions, fenarimol is used in mixture with other fungicides against cereal diseases in preference to nuarimol [45].

Examples for non-mildew diseases where the broad-spectrum fungicide fenarimol is used are leaf spot (Cercospora spp.) on sugar beets, Coccomyces spp. on cherries, and Ustilago-, Sclerotinia- and Fusarium-diseases on turf [46].

In vitro-tests with *nuarimol* [40, 47] revealed a strong inhibition of mycelial growth at ED 50-values often below 1 µg/ml. Fungi especially sensitive to the compound were in the genera Monilinia, Septoria, Alternaria, Fusarium, Rhizoctonia and Ustilago. Thus, a wide range of Ascomycetes, Fungi imperfecti and Basidiomycetes are inhibited by the compound.

In agriculture, nuarimol, known under the trade names Trimidal®, Triminol®, Murox® etc., is effectively used as a foliar fungicide in cereals against powdery mildew (Erysiphe graminis) in barley and wheat, and leaf blotch (Rhynchosporium secalis) in barley [48, 49, 50]. The performance of nuarimol against cereal rusts (Puccinia hordei, Puccinia recondita) is incomplete [48]. Typical application rates are 45–67.5 g a.i./ha.

In non-cereal indications, nuarimol is effective against Cercospora beticola in sugar beets and Cercospora spp. in bananas.

When applied as a seed treatment in cereals at 11–22 g a.i./100 kg seed nuarimol is recommended against a series of seed- and soil borne pathogens such as bunt and smut diseases (Tilletia spp.; Ustilago spp.), Fusarium (Gerlachia) nivale, and Pyrenophora graminea [49, 50].

Since nuarimol is taken up systemically by the roots, protection against early powdery mildew infections in barley is an additional beneficial effect of the seed treatment application [49, 50].

2.3 Imidazoles

Information on the broad in vitro-spectrum of *imazalil* was given in detail by Buchenauer and Röhner [40, 51]. Imazalil is used in agriculture in three different fields: as a seed treatment, as a foliar spray in vegetables, ornamentals and bananas, and as a post harvest application in citrus-, apple- and banana fruits.

As a cereal seed treatment, imazalil is used in combination with other fungicides or alone as an effective fungicide against Pyrenophora-, Fusarium- and Septoria species [52]. Its effect on Pyrenophora graminea at only 4–8 g a.i./100 kg seed is exceptionally good. In seed potatoes, imazalil is used against several fungal storage diseases such as gangrene, caused by Phoma exigua v. foveata, and skinspot caused by Polyscytalum pustulans [52, 53].

In foliar spray applications, imazalil is used against powdery mildews in cucurbits, roses, ornamentals and peppers as well as against grey mold (Botrytis cinerea), stem and fruit rot (Didymella spp.; Mycosphaerella spp.) and black spot on roses. An additional field of use is against Sigatoka-diseases on bananas caused by Mycosphaerella spp. [54].

A wide range of mold and rot-diseases are controlled by post-harvest applications in citrus fruits, bananas, apples, pears and pineapples.

Prochloraz is a new broad-spectrum fungicide which has good activity against several Ascomycetes and Fungi imperfecti but it is less active against Basidiomycetes [55].

Among the SBI-fungicides, prochloraz is outstanding because of its good effect against eyespot (Pseudocercosporella herpotrichoides) in cereals. For this indication, the compound is normally used at 480 g a.i./ha. In some countries, higher doses (750 g a.i./ha) or combinations with carbendazim are however, recommended [56,57]. In cereals, in addition, prochloraz shows effects against net blotch (Pyrenophora teres), leaf blotch (Rhynchosporium secalis), sharp eyespot (Rhizoctonia solani), Septoria-diseases and powdery mildew (Erysiphe graminis).

In fruit trees, control of scab (Venturia inequalis) and powdery mildew (Podosphaera leucotricha) on apple [58,59], and Monilinia fructicola on peaches [60] was reported.

Prochloraz has been tested in Japan against Pyricularia oryzae, Gibberella fujikuroi and Cochliobolus miyabeanus in rice and is under investigation as a cereal seed treatment in Europe.

Martin and Edgington [61], who tested the in vitro activity of the experimental fungicide *fenapanil* with fungi from all systematic classes, have found that its spectrum closely conforms to that of other azole compounds such as triadimefon, imazalil and fluotrimazole.

In orchards, fenapanil shows control of apple scab (Venturia inequalis), cedar-apple rust (Gymnosporangium juniperi-virginianae) and powdery mildew at application rates from 300–618 µg a.i./ml [42,59,62]. In cereals, most reports deal with the effectiveness of fenapanil as a seed treatment against soil- and seed borne diseases such as common bunt [63], Pyrenophora graminea and Urocystis occulta [64].

2.4 Triazoles

Fluotrimazol, the first triazole-fungizide which entered the market, is characterized by its narrow spectrum of fungicidal activity, this being restricted to the powdery mildew fungi [40,47,65]. In cereals, fluotrimazol, which is marketed under the trade name Persulon®, controls powdery mildew (Erysiphe graminis) at 125–187.5 g a.i./ha.

In fruits, vegetables, roses, grapes and others, a wide range of powdery mildews from the genera Podosphaera, Sphaerotheca, Erysiphe, Leveillula and Oidium can be controlled by fluotrimazol applied as a foliar spray at 0.0016 to 0.0125% a.i. In spite of its good and selective activity, which would allow its use in integrated pest control programs, the importance of Fluotrimazol has remained limited. Broadspectrum systemic SBI's with comparable mildew efficacy have been more successful.

Triadimefon, the first systemic triazole-fungicide with a broad spectrum of possible applications was introduced in 1973 by Bayer [65,66] and rapidly became known under its trade name Bayleton®. The in vitro-spectrum of triadimefon includes Ascomycetes, Basidiomycetes and Fungi imperfecti and is generally similar to the spectra of imazalil and nuarimol, although its effect against several Fungi imperfecti is less pronounced [40,47,67].

The high protective and curative effectiveness against the obligate parasites powdery mildew [68,69] and rust [70,71], together with a good control of other pathogens such as Rhynchosporium secalis [72,73], Septoria tritici [74], and Typhula incarnata [75,76] characterizes the performance of triadimefon in cereals. 1–2 treatments at 125–250 g a.i./ha are recommended; the higher amount is needed in cases of heavy rust infections.

In fruits, vegetables, grapes and ornamentals a wide range of diseases can be controlled at 0.0025–0.05% a.i. Some examples are: powdery mildew and rust in apples [77,78], powdery mildew of grapevine [79,80], coffee rust (Hemileia vastatrix) [81,82], Ceratocystis paradoxa in sugar cane and pineapple [83,84], and powdery mildew in cucurbits [85].

In plants and in many fungi, triadimefon is rapidly reduced to its corresponding secondary alcohol triadimenol [86,87]. In wheat, about 60% of the active ingredient which penetrates the leaves undergoes this reduction within 48 h at room temperature [88]. Because triadimenol is the more active compound in in vitro assays, Gasztony and Josepovits [89,90] regarded the enzymatic reduction of triadimefon to triadimenol as an activation. Of importance in this context is that the exact basis for the fungicidal activity of triadimefon is difficult to demonstrate since the fungi themselves often form at least small amounts of triadimenol when treated with triadimefon.

The technical active ingredient of triadimefon is a racemic mixture of (R)- and (S)-stereoisomers whose biological activities are similar [91].

Since *triadimenol*, the active ingredient of the fungicides Baytan® and Bayfidan®, can be formed from triadimefon in plants and fungi, the spectrum of activity of triadimefon and triadimenol are essentially the same [40,47].

Introduced in 1977 [92], triadimenol was initially and exclusively developed as a seed treatment in cereals. Due to its broad fungicidal spectrum and its systemic uptake into plants, triadimenol proved to be active against a wide range of cereal seedling pathogens at a rate of usually 15–37.5 g a.i./100 kg seed [93,94]. Of the seed-born diseases, Tilletia caries and T. foetida as well as Urocystis- and Ustilago-species are effectively controlled, even at a rate of 10 g a.i./100 kg seed. Other seed-and soil-borne infections which triadimenol controls are Pyrenophora diseases (P. teres, net blotch; P. avenae, oat leaf blotch), Septoria nodorum, (seedling blight), Typhula incarnata, (root rot) and Rhynchosporium secalis, (leaf blotch). The performance of the compound against Pyrenophora graminea (leaf stripe) is uncertain. For complete control of this disease, the addition of imazalil is necessary. Triadimenol has good activity against Fusarium (Gerlachia) nivale (snow mold). In cases of heavy infection pressure, however, the addition of fuberidazol results in a further improvement of control.

In addition to the spectrum already mentioned, an important attribute of triadimenol is that when it is applied as a seed treatment, it controls the air borne seedling diseases, powdery mildew (Erysiphe graminis) and rust (Puccinia spp.) [93,94].

Since triadimenol showed advantages over triadimefon in curative and eradicative treatments against powdery mildews and rusts, it has been further developed as a foliar fungicide under the trade name Bayfidan®. The spectrum of fungicidal activity and the application rates of this new product are similar to those of Bayleton®.

The reduction of triadimefon to triadimenol introduces a second chiral C-carbon into the molecule of the fungicide. This creates two diastereomeric forms, triadimenol A (= I or threo) and triadimenol B (= II or erythro). Each of these consists of two enantiomeric forms. Hence, triadimenol A includes the enantiomers A(—) (with the configuration 1S, 2R) and A(+) (1R, 2S) while triadimenol B is a racemic mixture of the enantiomers B(—) (1S,2S) and B(+) (1R,2R).

The biological properties of the two diastereomeric forms, and more recently the activity of the enantiomers have been studied intensively by various groups.

From earlier work it became known that by the reduction of triadimefon different fungal and plant species produce different diasteromeric mixtures of triadimenol in different amounts [40,86,95,96,97]. Since the diastereomers exhibit different fungitoxicity — triadimenol A being the more active isomer — attempts have been made to demonstrate a correlation between the degree of sensitivity of a fungus to triadimefon to the diastereomeric ratio of triadimenol formed from triadimefon by the fungus [90,87].

Krämer et al. [91] reported that the order of fungicidal activity of the enantiomers is A(—) > B(+) ≫ A(+), B(—). Recently Deas et al. [98,99] reexamined the fungal reduction of triadimefon on the enantiomeric level. These authors demonstrated that the amount and the pattern of reduction of triadimefon to triadimenol gives an explanation for the high sensitivity of some fungi to triadimefon. It completely fails to explain the low sensitivity of species such as Fusarium culmorum. From these data Deas et al. concluded that a simple relationship between sensitivity to triadimefon and preferential metabolic production of the most active A(—) enantiomer does not always exist.

Bitertanol, introduced by Bayer in 1979 [100], is structurally related to triadimefon. In comparison to these compounds, the much higher lipophilicity of bitertanol alters its usable fungicidal spectrum and its systemic mobility.

In vitro, bitertanol has a fungicidal spectrum which is comparable to that of triadimefon, however, its activity against several imperfect fungi such as Cercospora sp., Fusarium spp. and Cladosporium spp. [101] is more pronounced.

As a foliar fungicide (trade name: Baycor®), bitertanol is used in apples and pears — partially in mixture with conventional fungicides — as a protective and curative fungicide against scab (Venturia spp.) [102] and some other diseases such as rust (Gymnosporangium spp.), sooty blotch (Gloedes pomigena) and flyspeck (Schizotyrium pomi) at 0.005–0.025% a.i. [103].

In stone fruits, bitertanol controls Monilinia-diseases as well as some leaf spot pathogens such as shot-hole (Stigmina carpophila) [102]. The Sigatoka-complex (Mycophaerella spp.) and freckle disease (Guignardia musae), which is of importance in asiatic banana cultures, are treated in combination with oil at 0.125–0.15 kg a.i./ha [104]. From the multitude of other areas of use, the effects of bitertanol on Mycosphaerella-diseases, rust (Puccinia arachidis) and leaf blotch (Phoma arachidicola) in peanuts [104], leaf spot (Cercospora beticola) in sugar beets, leaf spot (Pseudopeziza ribis) and American gooseberry mildew (Sphaerotheca mors-uvae) in gooseberries and black currants [105], bean rust (Uromyces appendiculatus [106], and blister blight (Exobasidium vexans) in tea can be mentioned.

As a seed treatment in cereals, bitertanol — under the trade name Sibutol® — has a further area of application. Doses of 10–20 g a.i./100 kg seed are sufficient to control some bunts (Tilletia caries, Tilletia foetida) and covered smuts (Ustilago hordei, Ustilago avenae, Urocystis occulta). At higher application rates, pathogens such as snow mold (Fusarium nivale), seedling blight (Septoria nodorum) and the loose smuts (Ustilago spp.) are controlled. The effect of bitertanol (at 75 g a.i./100 kg seed) against the soil born pathogen of wheat, dwarf bunt (Tilletia controversa), must be mentioned [107]. This disease is difficult to control with other fungicides.

As with the related compound triadimenol, the active ingredient of bitertanol consists of four enantiomeric forms. Similar to triadimenol, experiments involving the bean rust fungus revealed a relationship between stereochemical configuration and fungicidal activity of the enantiomers. The, 1S, 2R enantiomer is more active in this test system than the other 3 enantiomers [108].

Propicanozole and etaconazole, both discovered by Janssen and developed by Ciba-Geigy, are two broad-spectrum fungicides which are not only structurally but also in their basic biological properties closely related. According to Urech et al. [68], propiconazole shows superior activity on cereals, whereas etaconazole is more suited for the control of diseases of deciduous fruit.

Basic information on the performance of *propiconazole* (trade names: Tilt® or Desmel®) in cereals is given by Urech et al. [109], Urech and Speich [110] and Smith and Speich [111]. The most important diseases effectively controlled by propiconazole, generally at 125 g a.i./ha (1–2 applications), are: Erysiphe graminis, Puccinia spp., Septoria tritici, and Rhynchosporium secalis. A slight to moderate control of Septoria nodorum infections has also been observed. Two treatments with propiconazole at 125 g a.i./ha are needed to control Pyrenophora teres (net blotch) in barley. Propiconazole is only partially effective against eyespot (Pseudocercosporella herpotrichoides) so that mixtures with carbendazim are recommended.

The broad spectrum of propiconazole becomes obvious from the wide range of non-cereal diseases where the compound has shown promising results at low application rates:

Cercospora leaf spot (Mycosphaerella spec.) and Sclerotium rolfsii are controlled in peanuts [112] while Cladosporium caryigenum and other leaf spot pathogens are controlled in pecans [113].

In rice, activity against five diseases has been reported: these are narrow brown leaf spot (Cercospora oryzae); brown leaf spot (Bipolaris oryzae); leaf smut (Entyloma oryzae); sheath spot (Rhizoctonia oryzae); and brown blotch (C. oryzae) [114].

In soybeans, propiconazole is known to be active against several foliar, pod and stem diseases; and in bananas, good results have been reported by Allen & Peasly [115] and by Mourichon and Beugnon [116] against Mycosphaerella-diseases.

Etaconazole, the second new compound which Ciba-Geigy presented in 1979, is also a broad spectrum fungicide which is mainly used in fruits. Staub et al. [117] gave some information on the in vitro activity against members of the Ascomycetes, Basidiomycetes and the Fungi imperfecti. Etaconazole proved to be a highly active compound which inhibited mycelial growth at concentrations as low as 0.1 ppm.

Several reports describe the protective and curative control of powdery mildew (Podosphaera leucotricha) and scab (Venturia inaequalis) in apple (a.o. [58, 59, 42]) and scab (Venturia pirina) in pears [59] by etaconazole at only 19–30 µg a.i./ml. According to Szkolnik [118] and Gilpatrick [119], etaconazole also gave control of cedar-apple rust (Gymnosporangium juniperi-virginianae), cherry leaf spot (Coccomyces hiemalis), and brown rot and also blossom blight of peach (Monilinia fructicola).

Sour rot, caused by Geotrichum candidum, and green mould on citrus fruits, two important post harvest diseases, are effectively controlled by etaconazole [120, 121].

Peanuts, bananas, beans, cucurbits and tomatoes are further crops in which etaconazole is effective against a variety of other diseases.

Because of the presence of two asymmetric carbon atoms in the dioxolane ring etaconazole is composed of four stereoisomeric forms. Information on the fungitoxic activity of the individual forms was given by Vogel et al. [122]. In vitro tests with two fungi showed a high activity of the 2R,4R-isomer against both. The fungicidal effect of the 2R,4S-form ranked high against Penicillium digitatum, but low against Botrytis cinerea. With the 2S,4S-isomer, the results were nearly inverse. The 2R,4R-form was consistently the weakest one.

Greenhouse tests with different host-parasite-systems revealed an order of activity as follows: 2S,4R > 2R,4S, 2S,4S > 2R,4R. In these tests, the 2S,4R-isomer was roughly 100 times more active than the weakest form (2R,4R).

Penconazole, with the trade name Topas®, also known as CGA 71818, is a new promising fungicide discovered by Janssen Pharmaceutica and developed by Ciba-Geigy.

According to Eberle et al. [123], penconazole has a striking activity against powdery mildews in pome fruits and grapes. Against scab (Venturia inequalis) in apples, the compound has a good curative action. However, mixture with a residual partner like captan is recommended because penconazole shows some weakness in protective activity. Eberle et al. [123] reported a good control of black rot (Guignardia bildwelli) on grapes and some secondary pome fruit diseases.

Diclobutrazol, introduced by ICI in 1979 [124], resembles triadimefon and triadimenol in structure as well as in its spectrum of fungicidal activity. Powdery mildew and especially rust diseases are controlled. V. Zitzewitz and Heckele [124] gave special attention to the effectiveness of the compound against yellow rust (Puccinia striiformis). They also found a good control of Rhynchosporium secalis (leaf blotch) and Typhula incarnata (snow rot). The good activity of diclobutrazol against rusts was confirmed by Javed [126] who reported effective control of coffee leaf rust (Hemileia vastatrix).

Although diclobutrazol, like triadimenol, contains two chiral C-carbons, only one diastereomere, the threo-form (2R,3R and 2S,3S), is included in the commercial product. Baldwin and Higgins [127] reported that from this diastereomere, the 2R,3R-enantiomer is responsible for almost all of the fungicidal activity.

The current literature indicates that *flutriafol* (Code-No.: PP 450), which was presented by ICI in 1983 [128, 129], is a promising fungicide which can be used as a foliar spray or as a seed treatment for cereals. At 125 g a.i./ha, control of Erysiphe graminis, Puccinia spp., Septoria spp., Helminthosporium spp., and Rhynchosporium secalis in barley and wheat has been reported.

As a seed treatment, flutriafol provides good control of soil- and seed-borne diseases such as Tilletia caries and Ustilago nuda, at 7.5 g a.i./100 kg seed; against foliar diseases such as powdery mildew, it is effective at 30 g/100 kg seed. As with most other SBI-fungicides, flutriafol also seems to require the addition of carbendazim for complete control of eyespot (Pseudocercosporella herpotrichoides).

In regard to *S-3308*, which is a new triazole fungicide presented by Sumitomo Chemical Co. in 1983, little information has been published. Funaki et al. [130] mention that it has good preventive, curative and partially systemic activity against powdery mildews, rusts, scabs and other diseases. Tests against Monilinia-diseases in peaches and cherries have also become known.

Another new triazole compound was presented in 1983 by BASF under the code number *BAS 454 06 F* [131]. The product has been broadly tested in cereals (Erysiphe graminis, Septoria nodorum), pome fruit (Venturia spp.), sugar beets (Cercospora beticola, Erysiphe betae), coffee (Hemileia vastatrix) and bananas (Mycosphaerella spp.).

PP 969 is a new and interesting compound with uncommon systemic properties from ICI. The compound was found to have excellent effects against coffee rust (Hemileia vastatrix), Sigatoka (Mycosphaerella musicola) on banana plants, and powdery mildew and leaf scab on apple when applied using unusual methods such as injection into the trunk, or stem- or soil treatment. When sprayed directly onto the leaves, the fungicidal effect was weak [132].

In 1984 Du Pont presented a new SBI-fungicide under the code number *DPX H6573*. According to Fort and Moberg [133], DPX H6573 provides control of a broad spectrum of diseases. The examples given by these authors demonstrated activity of the compound against eyespot (Pseudocercosporella herpotrichoides), rust (Puccinia spec.), powdery mildew (Erysiphe graminis), glume blotch (Septoria nodorum) and leaf blotch (Phynchosporium secalis) in cereals. In apples, scab (Venturia inaequalis) and powdery mildew (Podosphaera leucotricha) and in grapes, grape powdery mildew (Uncinula necator) were controlled. Finally, acceptable activity against Cercospora leafspots on peanuts and sugarbeet was reported.

2.5 Morpholines

Dodemorph, the first morpholine-fungicide, was introduced in 1967. The compound is of limited importance, which is probably partially due to its narrow spectrum, this being restricted to powdery mildews. Currently, dodemorph or its acetate-salt is used as a foliar spray at a concentration of ca. 0.1% a.i. in ornamentals such as roses and chrysanthemum [134, 135].

Tridemorph, introduced in 1969 by BASF, was — together with Triforine — the first sterol-biosynthesis inhibiting fungicide which became of greater importance. Tridemorph is particulary effective against powdery mildew in barley, but has less effect against species of rust fungi and powdery mildew in wheat [136]. The application rate is usually 560 g a.i./ha.

Hess and Heimes [137] gave a detailed summary of the possible uses of Calixin®, the trade name of tridemorph, in non-cereal crops. In addition to use on several powdery mildew fungi in the genera Sphaerotheca (e.g. in cucumbers and roses), Microsphaera (oak), Oidium (Hevea) and Uncinula (grapevine), the compound has been proposed for use against a wide range of Ascomycetes, Basidiomycetes and Fungi imperfecti. Thus diseases such as Yellow and Black Sigatoka and Black leaf streak in bananas can be controlled. The compound is also useful against tea blister blight (Exobasidium vexans) and pink disease (Corticium salmonicolor) in rubber.

The dodecyl-analog of tridemorph, *aldimorph*, was developed as a fungicide by

VEB Fahlberg-List in the German Democratic Republic. Under its trade name Falimorph®, the compound is mainly recommended for controlling barley powdery mildew; in order to achieve fungicidal control comparable to tridemorph, higher application rates are necessary [138].

Fenpropimorph, introduced in 1979 by Maag and BASF [136,139], is the most recent morpholine fungicide. The material became rapidly known under its trade names Corbel® and Mistral®.

Currently, fenpropimorph is exclusively used as a cereal fungicide which has interesting protective and eradicative properties, especially against powdery mildew (Erysiphe graminis) in wheat and barley. Bohnen and Pfiffer [139] pointed out that in addition to being active against mildew, good control of cereal rusts (Puccinia spp.) and Rhynchosporium secalis have been found. The recommended rate against all these diseases is 750 g a.i./ha.

Since fenpropimorph is not sufficiently effective against Septoria spp., net blotch (Pyrenophora teres) and eyespot (Pseudocercosporella herpotrichoides), mixtures with Chlorothalonil or Carbendazim, respectively, are necessary to control these pathogens [140].

Although appropriate formulations are not yet commercially available, Bohnen and Pfiffner [139] reported good control of Rhizoctonia-, Tilletia-, Ustilago-, and Helminthosporium-diseases when fenpropimorph is used as a cereal seed dressing.

Questions concerning the fungicidal activity of sterically different morpholine-derivaties have been discussed in detail by Pommer [141]. With the tridecyl- and the N-cyclododecyl-2,6-dimethylmorpholines, which include tridemorph and dodemorph, respectively, no significant difference was detectable between corresponding cis- and trans-forms. In the case of the 3-phenylpropylamines, which include the fungicide fenpropimorph, the cis-2,6-dimethylmorpholine derivative shows considerably better fungicidal activity than the corresponding trans-form. Consequently, the pure cis-compound has been developed under the common name fenpropimorph. This molecule has a chirality center at the intermediate propane chain, so that fenpropimorph contains 2 enantiomers. According to Pommer [141] the laevorotatory enantiomer (configuration S(—)) is more active against powdery mildew and rust than the dextrorotatory form (R(+)).

2.6 Table of Interesting SBI-fungicides (Overview)

The following table gives an overview of some of the important biological, physical and toxicological parameters of the SBI-fungicides.

Physical and toxicological data were mainly taken from Perkow [142]. In some cases, data from recent publications (cited in Chapter 2) of the company which developed the fungicide were used.

For new compounds, the data were taken from publications which are cited in Chapter 2.

Structure	1) Common name[a] 2) Trade name (s) 3) Experimental No.	1) Toxicity[b] 2) Solubility[c] 3) Vapour pr.[d] 4) Company 5) Year	Typical or proposed uses 1) Foliar application 2) Seed treatment 3) Remarks
Piperazines 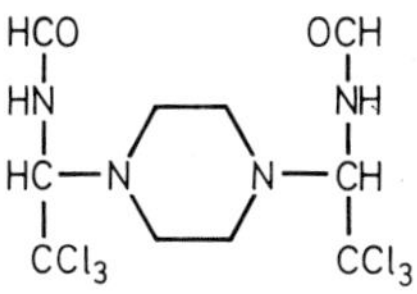	1) Triforine 2) Saprol® Funginex® 3) CELA W 254	1) >16000 2) 30 3) $2{,}6 \times 10^{-7}$ 4) Celamerck 5) 1971	1) Powdery mildew, rust in fruits, vegetables, ornamentals, cereals; Monilinia, scab in fruit trees; leaf spot diseases in vegetables; mummy berry disease in blueberries 2) Powdery mildew in barley
Pyridines 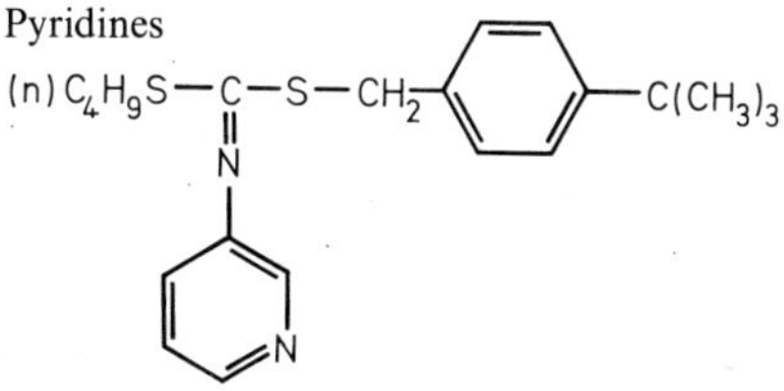	1) Buthiobate 2) Denmert® 3) S-1358	1) 2700–4900 2) 0.96[e] 3) 6×10^{-7} 4) Sumitomo 5) 1975	1) Powdery mildew in vegetables, fruits and ornamentals 3) Locosystemic
Pyrimidines 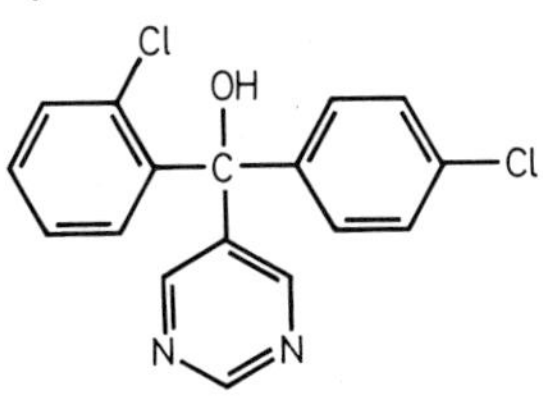	1) Fenarimol 2) Rubigan® Rimidin® 3) EL 222	1) 2500 2) 13.7[f] 3) $<1.3 \times 10^{-7}$[f] 4) Eli Lilly 5) 1975	1) Scab in pomefruits, powdery mildew in fruits, grapes, vegetables
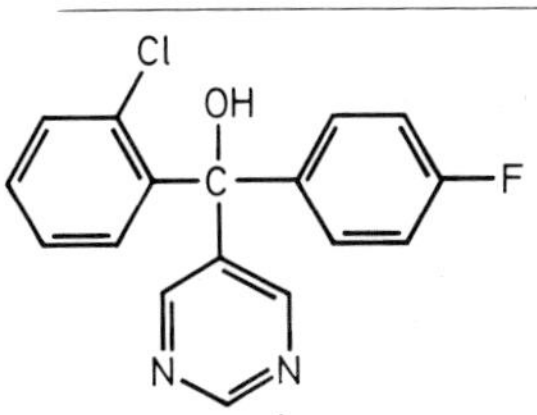	1) Nuarimol 2) Trimidal® Trimunol® Murox® a.o. 3) EL 228	1) 1250 ♂ 2500 ♀ 2) 26[f] 3) $<2 \cdot 10^{8}$[e] 4) Eli Lilly 5) 1975	1) Powdery mildew, leaf blotch in cereals; leaf spot in sugar beets 2) Soil-, seed- and airborne diseases in cereals
Imidazoles 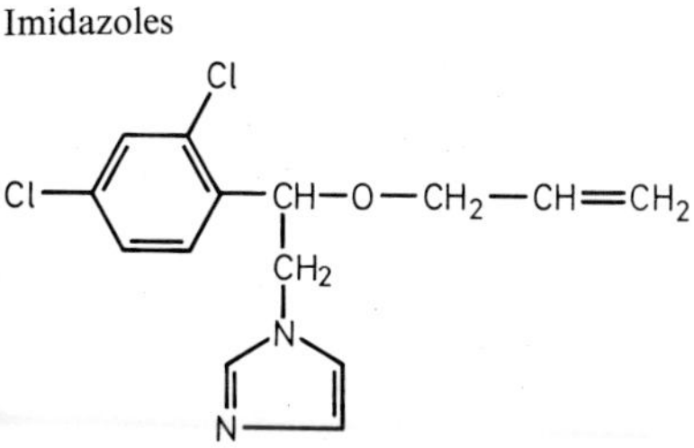	1) Imazalil 2) Fungaflor® Fungazil® Fecundal® 3) R 23 979 (free base) R 27 180 (sulphate salt)	1) 343–227 2) 293 3) 7×10^{-8} 4) Janssen 5) 1972	1) Powdery mildew, mold and rot diseases in vegetables, ornamentals; Sigatoka on bananas 2) Seed- and soilborne diseases (esp. leaf stripe) in cereals; storage diseases in potatoes 3) Post harvest fungicide (citrus, apple)

(Table continued)

Structure	1) Common name[a] 2) Trade name (s) 3) Experimental No.	1) Toxicity[b] 2) Solubility[c] 3) Vapour pr.[d] 4) Company 5) Year	Typical or proposed uses 1) Foliar application 2) Seed treatment
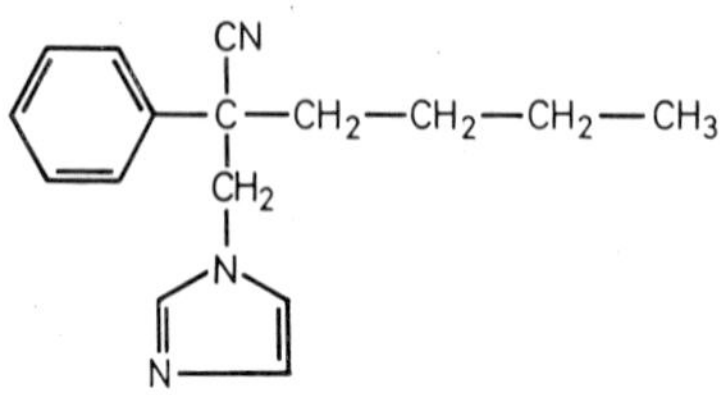	1) Fenapanil (Fenapronil)[h] 2) Sisthane® 3) RH 2161	1) 1590 2) 3) 4) Rohm & Haas 5) 1978	1) Scab, rust, powdery mildew in pomefruits
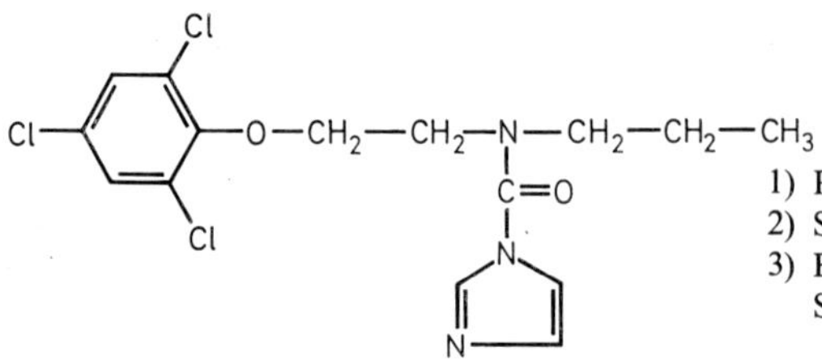	1) Prochloraz 2) Sportak® 3) BTS 40542 SN 80 109	1) 1600 2) 48 3) $0{,}8 \times 10^{-9}$ 4) FBC 5) 1977	1) Eyespot, powdery mildew, netblotch in cereals 2) Seed- and soilborne diseases in cereals 3) Locosystemic
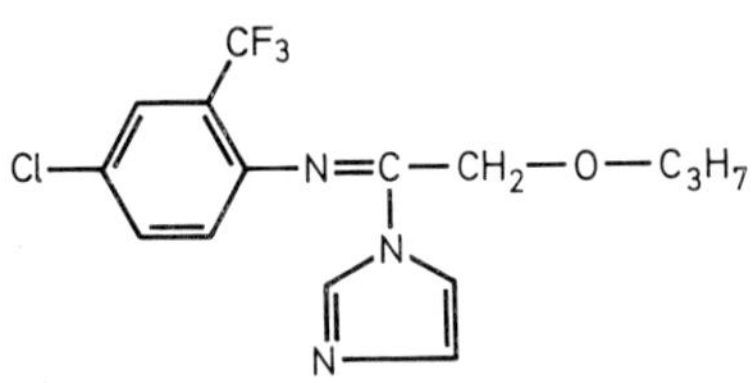	1) Triflumizole 2) 3) NF 114	1) 2) 3) 4) Nippon Soda 5) 1983	1) Powdery mildew, scab, leaf spot diseases in fruits, vegetables, ornamentals*
Triazoles 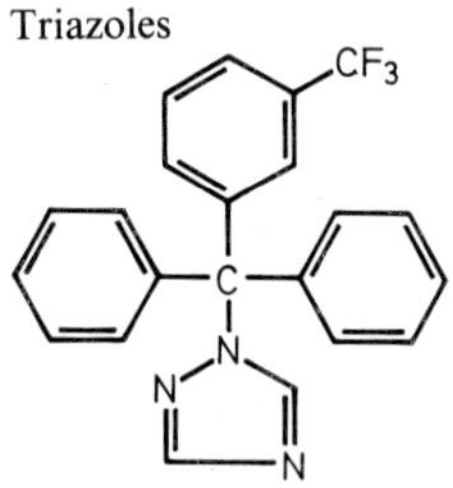	1) Fluotrimazole 2) Persulon® 3) BUE 0620	1) >5000 2) 0,5 3) $3{,}6 \times 10^{-7}$ 4) Bayer 5) 1973	1) Powdery mildew in fruits, ornamentals, vegetables 3) Locosystemic
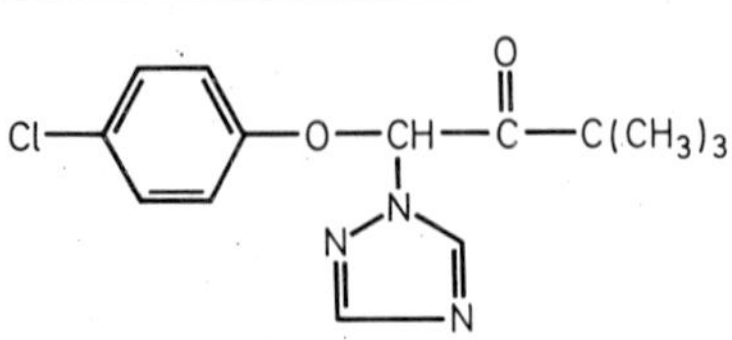	1) Triadimefon 2) Bayleton® 3) MEB 6447	1) 750–1200 2) 70 3) $<10^{-6}$ 4) Bayer 5) 1973	1) Powdery mildew, rust and other leaf diseases in cereals; powdery mildew, rust, a.o. in fruits, vegetables, grapes, ornamentals; coffee rust

(Table continued)

Structure	1) Common name[a] 2) Trade name (s) 3) Experimental No.	1) Toxicity[b] 2) Solubility[c] 3) Vapour pr.[d] 4) Company 5) Year	Typical or proposed uses 1) Foliar application 2) Seed treatment 3) Remarks
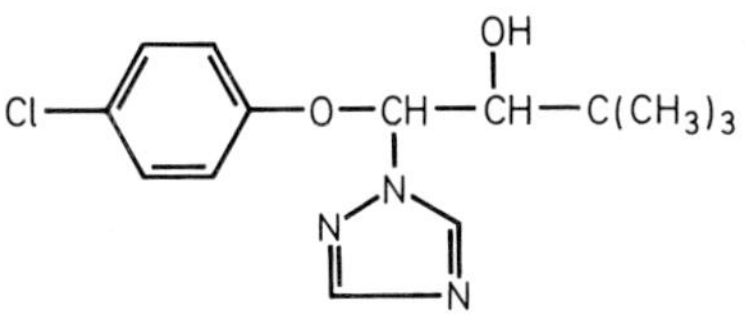	1) Triadimenol 2) Baytan® Bayfidan® 3) KWG 0519	1) 700–1500 2) 95 3) $\ll 10^{-6}$ 4) Bayer 5) 1977	1) Powdery mildew, rust and other leaf diseases in cereals; powdery mildew, rust, a.o. in fruits, vegetables, grapes 2) Soil-, seed- and airborne diseases in cereals
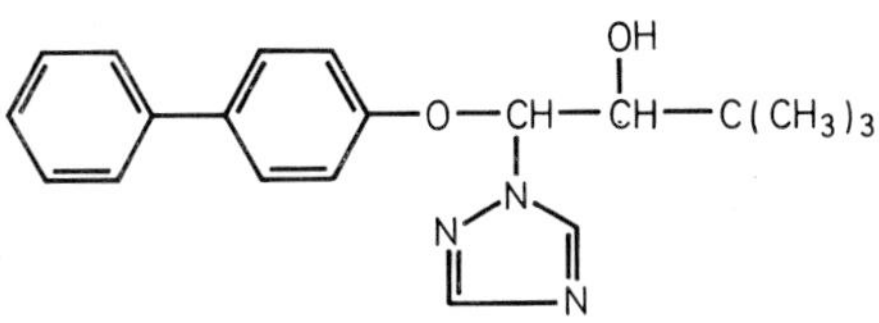	1) Bitertanol (Biloxazol)[h] 2) Baycor® Sibutol® 3) KWG 0599	1) 5000 2) 5 3) $\ll 10^{-6}$ 4) Bayer 5) 1978	1) Scab in pomefruits; Monilinia, leaf spot diseases in stone-fruits; rust, powdery mildew, leaf spot diseases in peamuts, vegetables, ornamentals; Sigatoka in bananas 2) Seed- and soilborne diseases in cereals 3) Locosystemic
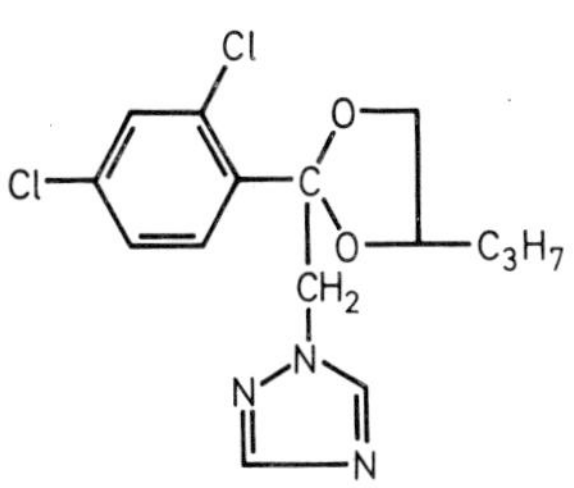	1) Propiconazole 2) Tilt® Desmel® Radar® 3) CGA 64 250	1) 1517 2) 110 3) $1{,}3 \cdot 10^{-6}$ 4) Janssen/ Ciba-Geigy 5) 1979	1) Powdery mildew, rust, netblotch and other leaf diseases in cereals; various fungal diseases in peanuts, soybeans, rice, bananas
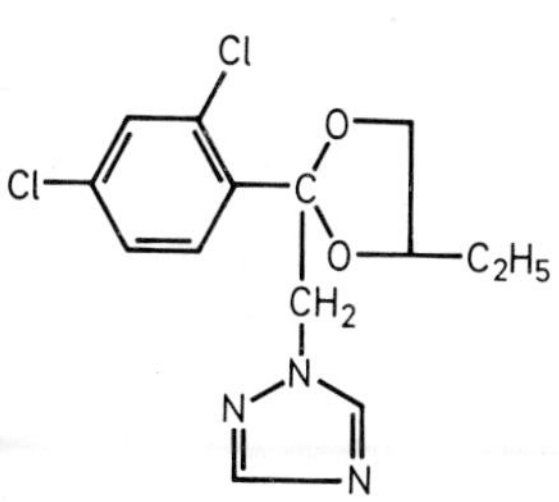	1) Etaconazole 2) Sonax® Vangard® Vanguard® 3) CGA 64 251	1) 1343 2) 80 3) 4) Janssen/ Ciba-Geigy 5) 1979	1) Powdery mildew, scab, rust, Monilinia, leaf spot in fruit trees; various fungal diseases in fruits and vegetables

Structure	1) Common name[a] 2) Trade name (s) 3) Experimental No.	1) Toxicity[b] 2) Solubility[c] 3) Vapour pr.[d] 4) Company 5) Year	Typical or proposed uses 1) Foliar application 2) Seed treatment 3) Remarks
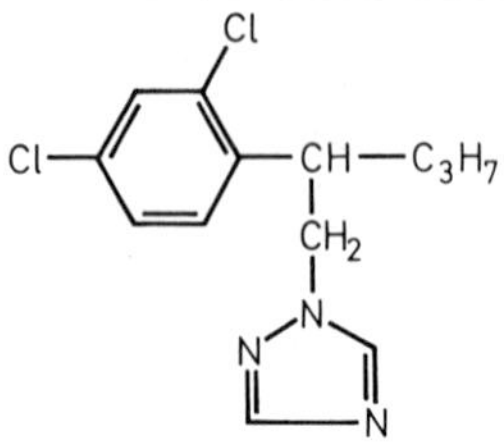	1) Penconazole 2) Topas® 3) CGA 71 818	1) 2125 2) 70 3) $2{,}1 \cdot 10^{-6}$ 4) Janssen/ Ciba-Geigy 5) 1983	1) Powdery mildew in pomefruits and grapes; scab in pomefruits*
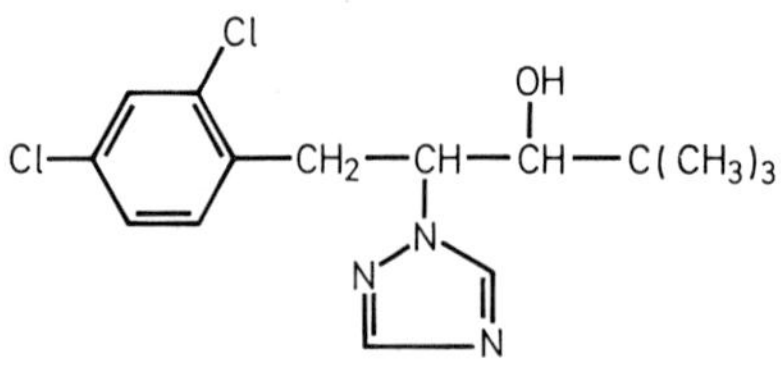	1) Diclobutrazol 2) Vigil® 3) ICI 70 180 F PP 296	1) ~4000 2) 9 3) $1{,}3\text{–}2{,}6 \cdot 10^{-8}$ 4) ICI 5) 1979	1) Rust, powdery mildew, and other diseases in cereals; coffee rust
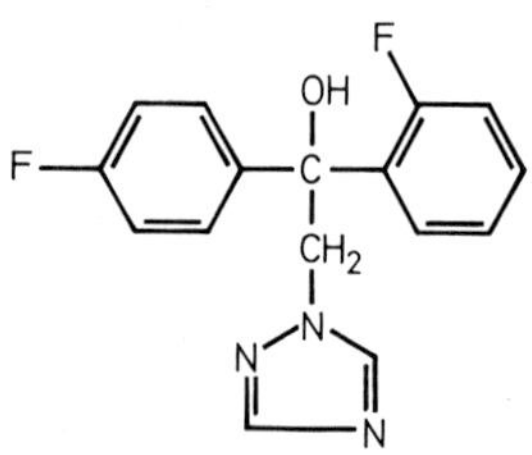	1) Flutriafol 2) Impact® 3) PP 450	1) 1140 ♂ 1480 ♀ 2) 130 3) $4 \cdot 10^{-9}$ 4) ICI 5) 1983	1) Powdery mildew, rust, and other diseases in cereals* 2) Seed-, soil- and airborne diseases in cereals*
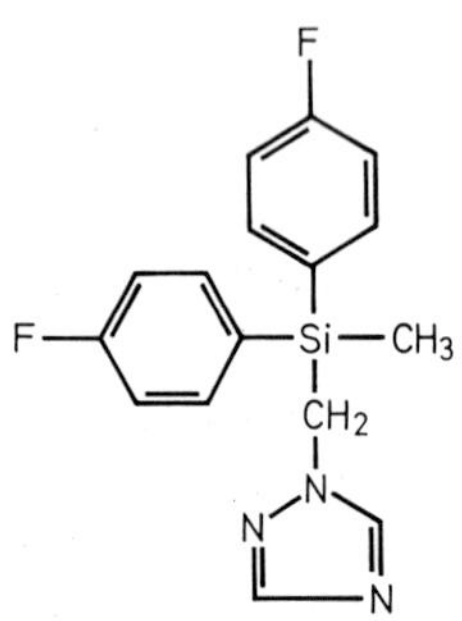	1) 2) 3) DPX H 6573	1) 1110 ♂ 674 ♀ 2) 3) 4) DuPont 5) 1984	1) Eyespot, rust, powdery mildew and other leaf diseases in cereals; scab, powdery mildew in apples; leaf spot in peanut, sugarbeet*
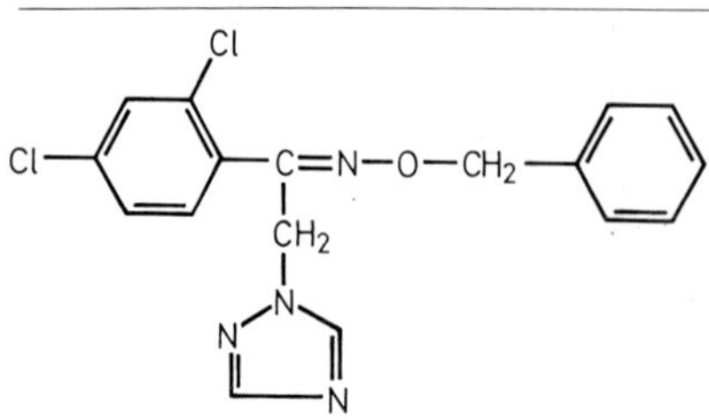	1) 2) 3) BAS 45 406 F	1) 4640 2) 3) 4) BASF 5) 1983	1) Powdery mildew, Cercospora spp., scab, a.o.*

Structure	1) Common name[a] 2) Trade name (s) 3) Experimental No.	1) Toxicity[b] 2) Solubility[c] 3) Vapour pr.[d] 4) Company 5) Year	Typical or proposed uses 1) Foliar application 2) Seed treatment 3) Remarks
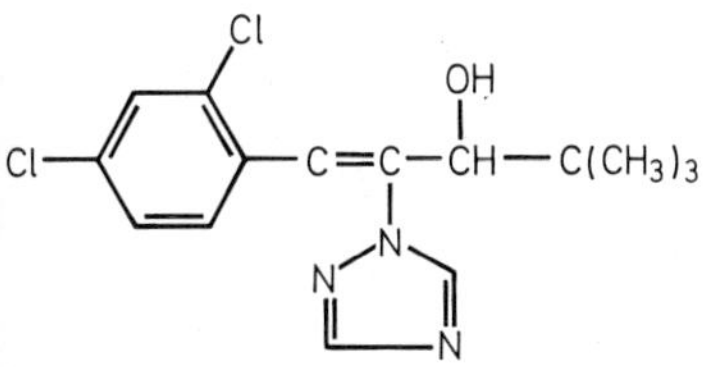	1) 2) 3) S-3308	1) 2) 3) 4) Sumitomo 5) 1983	1) Powdery mildew, rust, scab, Monilinia-diseases*
Morpholines 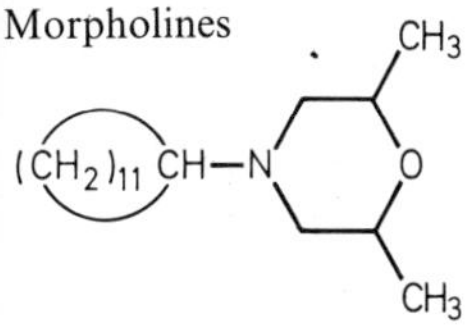	1) Dodemorph 2) Dodemorph® Meltatox® a.o. 3)	1) ~4800 2) <10 3) $2{,}5 \cdot 10^{-5}$ (free base) 4) 5) 1967	1) Powdery mildew in ornamentals
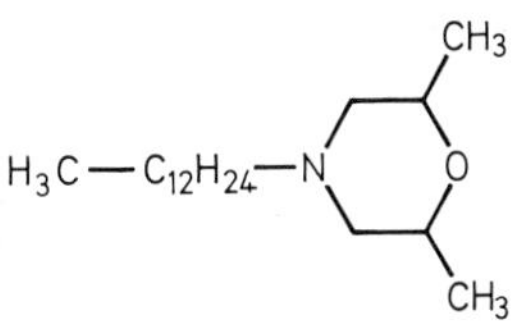	1) Tridemorph 2) Calixin® 3)	1) 1000[i] 2) 100 3) $4 \cdot 10^{-4}$ 4) BASF 5) 1969	1) Powdery mildew in cereals; Sigatoka in bananas; powdery mildew and other diseases in various crops
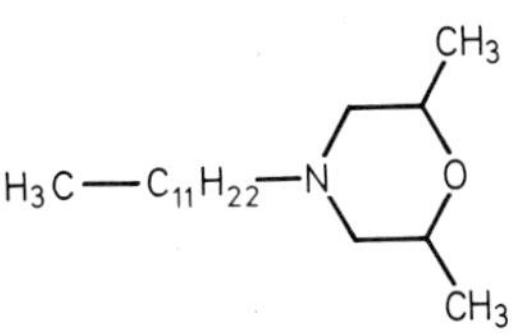	1) Aldimorph 2) Falimorph® 3)	1) 3500 2) 20[e] 3) $1{,}2 \cdot 10^{-3}$[g] 4) VEB Fahlberg-List 5) 1980	1) Powdery mildew in barley
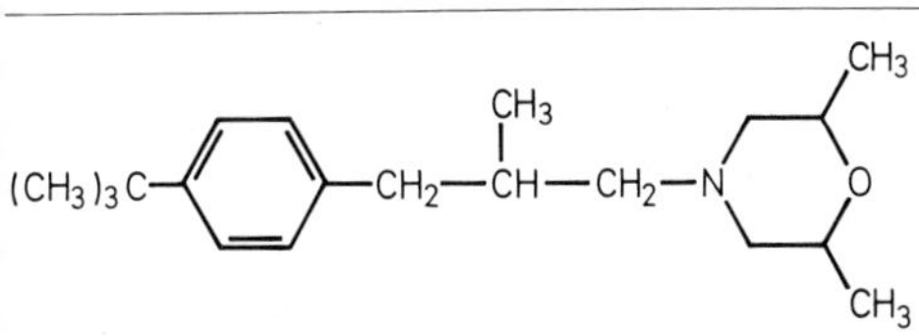	1) Fenpropimorph 2) Corbel® Mistral® 3) BAS 42 100 F	1) 3515 2) 6,8 3) $3.2 \cdot 10^{-5}$ 4) Maag/ BASF 5) 1979	1) Powdery mildew, rust, leaf blotch in cereals 2) Seed- and soil-borne diseases in cereals*

[a] ISO/BSI approved or proposed common name; [b] oral acute toxicity in rats (LD_{50} (mg/kg)); [c] solubility in water (mg/l) at usually 20 C; [d] vapour pressure (mbar) at usually 20 C; [e] determined at 25 C; [f] determined at 25 C; pH 7; [g] determined at 21.8 C; [h] former proposed common name; [i] LD_{50} (mm^3/kg) for the technically pure compound.
* Marks uses for which only preliminary data are available

3 Systemic Properties

3.1 General

The fact that most SBI-fungicides are able to penetrate into the plant is a major reason for the success of this type of fungicide. The heterogenicity of chemical structures in the SBI's has caused a great degree of variability in the capacity of the individual compounds to penetrate the outer barriers of the plant and eventually become distributed. Differences in the degree of systemic distribution should not be taken as a measure of efficacy. In practice, compounds with moderate or weak systemic activity can have clear advantages since they are not rapidly transported away from their side of action.

Generally, the ability to use systemic fungicides as a foliar spray has been regarded as a progressive step in the protection of plants against fungal diseases. There are three reasons for this:

1. The action against fungal diseases is often more complete since areas missed during the initial spray treatment are provided with protection. This is due to redistribution of the chemical in the plant. Since the SBI's are predominantely transported from the lower to the upper parts of the plant, protection is confined to such cases where a sufficient spray deposit has reached the lower parts of the leaves or shoots.
2. Since the surface deposits left by spray application are subject to weathering, penetration into the plant reduces the risk of total failure of disease control under unfavourable weather conditions.
3. The most important point is the curative effect, the possibility to control diseases which have already established in the plant. In many cases, application of fungicides can be limited to such cases where first infections are already visible, or where systems of prognosis are available which allow prediction of latent infections.

Edgington [143], who recently discussed the structural requirements of systemic fungicides, coined the terms "apoplastic" and "symplastic" for the patterns of movement of pesticides in the plant. The non-living part of the plant is the apoplast, i.e. xylem vessels, tracheids, cell walls and cuticle. Substances translocated in the apoplast move upward (or acropetally) with the evaporation stream. Prolonged uptake of substances which are predominantely transported in the apoplast thus results in characteristic accumulations in leaf margins or leaf tips.

The special term "locosystemic" is often used for molecules which are mobile only in a limited part of the apoplast, the cuticle and the adjacent cell walls. These are, therefore, not translocated over longer distances within the plant. Hence, pathogens which are growing near the plant surface (as for example apple scab and powdery mildew), can be controlled in curative treatments with locosystemic fungicides.

All currently known SBI-fungicides are at least partially mobile in the apoplast; their mobility in the symplast, which includes the living part of the plant (i.e. cytoplasm, plasmodesmata) is unfortunately thus far minimal. It is obvious from their physiological side-effects that the SBI-fungicides are able to cross the plasmalemma, however, they do not move in greater amounts in the phloem, the system in the plant which transports assimilates downward in the plant.

The vapour pressure of some SBI-fungicides is high enough to generate biologically effective concentrations in the gasphase at room temperature. Because of this, a fungicidal effect in plant parts situated beneath the area of application can often be observed. This transport of active ingredient via the gasphase sometimes simulates symplastic movement and can thus cause experimental errors.

The magnitude of the vapour pressure of an individual compound (see Chapt. 2.6) gives an impression of its potential concentration in the gasphase after spray application. However, to exhibit a fungicidal effect via the gasphase, additional factors like stability to light, rate of absorption by plant- or fungal surfaces and the degree of fungicidal potential seem to have an important influence.

3.2 Piperazines, Pyridines and Pyrimidines

Triforine, when applied to roots of barley and tomato seedlings, was readily taken up from aqueous solutions but less efficiently taken up from soil with high organic matter content [144, 145]. Von Bruchhausen and Stiasni [146] noted, that after soil drench application in sandy soil, barley shoots contained a maximum concentration of 7.5 ppm at the second day after treatment. The total uptake after 24 days was about 3% of the applied dosage.

Whereas apical translocation in the apoplast after root uptake is well documented, the mobility of triforine after leaf application seems to be rather limited.

Kato et al. [37] demonstrated vapour phase activity with *buthiobate*. The growth of cucumber mildew was prevented over the distance of about 16 mm beyound the edge of filter paper discs containing this fungicide. Ohkawa et al. [147] found that buthiobate was absorbed by the plant through leaves or roots, but translocation into other parts of the plant seemed to be limited to a very small extent.

Some of the systemic aspects of *fenarimol* were presented in detail by Brown and Hall [148]. These authors pointed out that in greenhouse trials fenarimol was rapidly absorbed by the leaves of bean, apple, cucumber, and wheat. By placing fenarimol-treated filter discs onto leaves of cucumbers, an acropetal movement in the leaf could be demonstrated. When aluminium foil discs separated the treated paper discs from the leaf surface, Brown and Hall noted a circular inhibition of powdery mildew caused by the vapour action of the fungicide. In contrast to this, Buchenauer [47], who compared the vapour phase activity of several fungicides, found that fenarimol had only a weak effect against barley powdery mildew. Root uptake of fenarimol was reported by Cardoso et al. [149] and by Buchenauer and Röhner [97].

Little quantitative data are available on the systemic activity of *nuarimol*, although several general observations indicate that the properties of nuarimol are comparable to those of Fenarimol. Cardoso et al. [149] mentioned that nuarimol was translocated into hypocotyls of soybeans which originated from fungicide-treated seeds. In addition to fenarimol, Buchenauer and Röhner [97] also found acropetal transport of nuarimol into barley plants after soil treatment. This is also demonstrated by the effect which the fungicide has on powdery mildew of seed-treated cereal seedlings.

3.3 Imidazoles

Corresponding to all other SBI's, *imazalil* is transported mainly in the apoplast in acropetal direction. Vonk and Dekhuizen [150] demonstrated that after seed treatment with 1.8 g imazalil/100 kg seed, about 6% of the active ingredient was present in the leaves of 3 weak old barley plants. Extraction of the leaves showed that they contained a maximum of 0.07 ppm imazalil. Reisdorf et al.[151] who investigated the effect of imazalil on barley seeds infected with leaf stripe (Pyrenophora graminea), supported these results. They found similar concentrations of the active ingredient in the shoot of the plants after root application. They did, however, suggest that these concentrations were too low to have a curative effect on the leaves infected with leaf stripe. They concluded, that the beneficial effect of imazalil against leaf stripe after seed dressing was mainly due to its effect on the inoculum mycelium in the pericarp of the seed kernel, and not to its systemic properties.

After treatment of first leaf tips of barley and cucumber, only small amounts of imazalil were found to be transported basipetally in the leaf or further down to the roots[52].

Prochloraz is apparently only poorly translocated in the plant[55]. This was confirmed by Cooke et al.[152] in experiments with strawberry. According to Schneider et al.[56], locosystemic activity, i.e. penetration into the plant without important translocation, is responsible for the curative activity of prochloraz.

Martin and Edgington[61,153] studied the systemic activity of *fenapanil.* When they applied the radiolabled fungicide to leaves of barley- and soybean plants, they found that 98.1 and 86.6% of translocated active ingredient were transported, respectively, in the apoplast in barley and soybean, 2 and 13.5% were transported, respectively, in the symplast. In bioassays, the authors partly confirmed these results; in barley, basipetal translocation could not be found using inhibition of powdery mildew as a means of definition.

Kelly and Jones[154], who worked with fenapanil in apple leaves, reported a foliar uptake of about 50% within 1–3 days. Fenapanil showed no gas-phase activity in these experiments.

3.4 Triazoles

Fluotrimazol is essentially a nonsystemic fungicide. Translocation from the roots into shoots of barley and wheat plants is only slight[47]. Since the fungicide has some curative activity, locosystemic movement in the plant is probable.

Triadimefon is one of the most intensively studied fungicides in regard to its systemic properties. Summarizing many plant studies which used either bioassays as an indirect proof of systemic transport[155,156,157,158,159,160,161] or analytical tools[86,89,90,162,163,164], it can be stated, that triadimefon (and/or the metabolite triadimenol) is rapidly taken up by the roots and the shoot surface and translocated mainly acropetally in the apoplast. Symplastic transport seems to be rather limited. Führ et al.[162], who used C-14-labelled triadimefon, found that only 0.3% of the applied radioactivity was translocated basipetally in barley plants.

In these experiments, upward translocation was 28–51% within 12 days. Brandes et al.[163], who worked with cucumber leaves, found that 4.7 and 0.5% of the applied radioactivity, respectively, were translocated acropetally and basipetally from the treated areas.

Kraus[164], who studied the speed of uptake of triadimefon into grape leaves, reported that only 40% of the applied active ingredient were removable with water 45 minutes after application to the leaf surface.

During experimental work with triadimefon, it became obvious that the compound has a marked activity via the vapour phase. Scheinpflug and Paul [165] reported 46 mm zones of inhibition of Erysiphe cichoracearum on cucumber after application of 1,25 μg on filter paper discs which were separated from the leaves by aluminium foil.

In densely growing cereal stands, the vapour phase activity has been found to contribute to the performance of a fungicide [159, 166]. In small plot experiments in which untreated reference plots are next to treated ones, errors in interpretation can occur [167].

Since *triadimenol* was first developed as a seed treatment, work on its systemic properties has been mainly restricted to its uptake and translocation after application as a seed dressing.

Steffens et al. [168] intensively studied the uptake and distribution of radiolabeled triadimenol in spring barley and spring wheat. According to these authors, up to 7.7% of the applied radioactivity was taken up by wheat seedling shoots within 51 days. After this time, the plants had reached growth stage 30–31 on Zadok's scale.

The main uptake into the plant took place during development of the first three leaves, i.e. during the first four weeks after sowing. In regard to the distribution of radioactivity in individual leaves, it was found that leaves No 1, 2 and 3 contained 3.9, 1.4 and 0.5% of the total activity, respectively. In spring barley, the concentrations in the first three leaves decreased even more rapidly, since the fresh weight of the three leaves increased more rapidly.

In contrast to the related compounds, triadimefon and triadimenol, *bitertanol* is a compound with limited systemic properties. This product, when applied as a foliar spray, is able to penetrate the leaf surface. This explains its curative action against many diseases [104]. Nevertheless, Scheinpflug und van den Boom [104] have demonstrated with peanut leaves that translocation of bitertanol to other parts of the plant is minimal. In accordance with its low vapour pressure (s. Chapt. 2.6) biological activity of bitertanol via the gas-phase is negligible.

The fungicide *propiconazole* is a highly systemic compound which is taken up by roots and leaves and translocated predominantly acropetally in the apoplast [110]. Its vapour phase activity seems to be as important as that of triadimefon [169], but detailed studies on this point are not currently available.

As with propiconazol, *etaconazole* is also taken up by plants and transported in the apoplast. Kelley and Jones [42], who studied the systemic properties of etaconazole in apple, reported that approximately 90% of the active ingredient was taken up within 12 h. These authors and also Szkolnik [58, 170] pointed out that etaconazole has a pronounced vapour activity. Szkolnik [170] obtained complete control of several powdery mildew diseases by applying etaconazole to cotton cloth in a greenhouse.

Diclobutrazol is a systemic fungicide which, according to Bent and Skidmore [124], shows translaminar movement and acropetal translocation from stem and leaf deposits. The authors reported a marked vapour action against powdery mildew over a distance of several centimeters.

Nevertheless, Rathwell and Skidmore [169], who compared the vapour activity of diclobutrazol with that of triadimefon and propiconazole, stated that diclobutrazol has no useful vapour action under field conditions. Buchenauer et al. [171], who applied diclobutrazol as a seed treatment, provided indirect proof of uptake by the roots of barley seedlings.

Three quite recent triazole fungicides deserve mention. *Penconazole*, which was studied by Eberle et al. [123], penetrates rapidly into plant tissues and is translocated acropetally. *Flutriafol*, the new compound from ICI, is transported in the apoplast as was shown by its effect against windborn seedling diseases after seed treatment [128]. Du Pont's new fungicide *DPX H6573* was reported to be a systemic compound which enters the plant rapidly [133].

3.5 Morpholines

Dodemorph is absorbed by leaves and translocated into non-treated parts of the plant. Uptake by roots has also been reported [134].

Investigations on the systemic behaviour of *tridemorph* made by Otto and Pommer [172] have proved that the active ingredient is rapidly taken up by barley leaves and transported acropetally in the plant. In cucumbers, tridemorph showed the same behaviour with a tendency to accumulate at the edges of the leaves. In banana plants, translocation was very slow.

Hamm [173], who investigated in detail penetration and transport of tridemorph in banana leaves, found a good penetration from the surface into the leaf tissue but a very weak acropetal or basipetal translocation.

Sharma and Sohi [174], who dealt with the systemic behaviour of tridemorph in cowpea (Vigna sinensis), and Vyas et al. [175] who did similar studies with peanuts, reported that after seed or soil treatment, there was sufficient fungicide in the leaves to control Rhizoctonia solani.

According to Bohnen and Pfiffner [139], *fenpropimorph* is rapidly taken up by roots and leaves of plants and translocated acropetally in the apoplast. The authors indicate that the amount of chemical transported basipetally is minimal. Fenpropimorph has a marked activity via the gasphase. Bohnen and Pfiffner [139] reported that powdery mildew-infected, untreated barley or wheat plants were protected when treated plants were placed 1 m away in greenhouses. Rathwell and Skidmore [169] also observed a strong vapour phase. They estimated that this was higher than the vapour activity of triadimefon and propiconazole.

4 Side Effects (Influence on Plant Physiology)

4.1 General

Research on side effects of pesticides on the physiology of plants is of importance as they may have a positive or negative effect on yield, an effect which compliments or negates control of the disease organism.

In the case of the SBI-fungicides, this reason is of special interest, since the plant-growth-regulatory (PGR) -side effect often dominates over their unspecific symptoms of phytotoxicity (e.g. necrotic flecks on leaves, withered leaf tips and leaf fall).

Consequently, from research with SBI-fungicides, some growth-regulators were detected which have been optimized for their PGR-potential. The plant growth-regulators ancymidol, which is related to the fungicide triarimol and paclobutrazol, closely related to diclobutrazol, are well known examples.

Obviously, in some classes of SBI's, there is a fluent transition from molecules with pronounced fungicidal activity and low PGR-potential to substances with the opposite properties. Stereo-isomers within one active ingredient often show clear differences in regard to their biological properties. For example, with the new triazole compound, S-3308, the R(—)-isomer has 100 times more fungicidal activity than the S(+)-isomer. On the other hand, the PGR-activities of the S(+)-isomer were more than 100 times stronger than those of the R(—)-isomer [130].

With paclobutrazol, the (S,S)-enantiomer showed the higher PGR-activity while the (R,R)-enantiomer had the higher fungicidal activity [176].

If one tries to classify the SBI-fungicides by virtue of their side-effects on plants, three groups can currently be distinguished.

The first group is characterized by typical growth-regulator effects on plants. It includes the pyrimidine-, imidazole-, and triazole compounds, although the affiliation of some newer compounds, such as prochloraz, is based only on structural analogy.

A second group includes the morpholine derivates. Due to their relatively small structural diversity it is presumed that they have a similar influence on plant physiology. Thus far, our knowledge of this group is insuffient: only one compound, tridemorph, has been intensively studied.

In view of their side effects, the last group of SBI-fungicides can only be negatively defined. This group includes members such as the the piperazine derivative triforine; based on their chemical structure or their macroscopically visible effects on plants these compounds do not fall into the either two categories.

4.2 Piperazines and Pyridines

Papers dealing with symptoms of plant incompatibility after *triforine* treatment in ornamentals [177, 178] or european larch [179] only describe unspecific visible symptoms of phytotoxicity; typical growth-regulator effects are not described.

In spring barley, treated at the recommended growth stages with triforine, Decallone et al. [180] could not detect significant changes in the grain protein fraction. Jalali and Domsch [181], however, found a general decrease in the amino acid content of root exudates of wheat plants after foliar application. A further side effect of triforine is the inhibition of germination of apple [182] and blueberry [183] pollen.

An interesting particularity of triforine is its marked phytotoxicity to certain lettuce cultivars, a crop on which triforine is not normally used. Inheritance studies showed that resistance to leaf damage caused by triforine is recessive and determined by a single gene [184, 185].

Literature on the possible side effects of the pyridine derivative *buthiobate* on plant physiology have not been published.

4.3 Pyrimidines, Imidazoles and Triazoles

Literature dealing with the effects of this group of SBI-fungicides (and the related growth-regulators) on the physiology of plants contains contradictory results. Thus far, it is not known if the effects caused by these chemicals are due to inhibition of gibberellic acid biosynthesis, inhibition of plant sterol synthesis, or both. Shive and Sisler [186], Coolbaugh and Hamilton [187], Sisler et al. [188] and Goldsmith et al. [189] investigated the biosynthesis of gibberellin in SBI-treated plants or in the gibberellin-producing fungus Gibberella fujikuroi. They found that SBI's of this group block the conversion of ent-kaurene to ent-kaurenoic acid in the biosynthesis of gibberellin. They confirmed these results by demonstrating that externally applied gibberellic acid could reverse the effects of the chemicals. Moreover Shive and Sisler [186], found that growth retardation of bean plants occured in the absence of net increase in sterol levels. They hypothesized, that changes of sterol levels in plants are hardly responsible for the growth retarding effects. In contrast to this, Schmitt and Benveniste [190], who used bramble cell cultures treated with the pyrimidine-derivative fenarimol, found a major change in sterol patterns, namely, an accumulation of 14α-methyl sterols and a nearly complete decrease in the normally dominating Δ^5-sterols of the plant.

Finally, Buchenauer and Röhner [191], who examined the effect of the triazole-fungicides triadimefon and triadimenol on barley seedlings, found interference with the plant sterol fraction and with gibberellin biosynthesis.

Apart from these biochemical changes, plants treated with SBI-fungicides of this group show — in many cases at doses well over those recommended in practice — quite uniform macroscopic symptoms. These symptoms — smaller, dark green leaves, shortening of internodes, retarded senescense of photosynthesizing parts of the plant — and their possible influence on physiological parameters like frost tolerance, resistance against dryness etc., will be described in more detail below.

Although many members of the pyrimidines, imidazoles and triazoles are known to induce typical growth regulator effects in plants it is so far difficult to compare the strength of the PGR-potential of individual compounds precisely. A rough evaluation of the growth retarding potential of some representive compounds on barley seedlings was given by Buchenauer [192].

At comparable concentrations, Buchenauer [192] found that seed treatment retarded the growth of 12 day old primary leaves (in comparison to untreated) as follows: fenapanil 8%, imazalil 15%, triadimenol 16%, propiconazole 20%, triadimefon 22%, nuarimol 23%, etaconazole 27% and diclobutrazol 28%. Studies in which different compounds were compared to determine their influence on parameters other than retardation of leaf growth in cereal seedlings are lacking.

Due to their structural relationship to the plant growth regulator ancymidol, the PGR-potentials of the pyrimidine fungicides *triarimol*, *fenarimol* and *nuarimol* were investigated soon after their introduction. In 1977 Buchenauer [47] reported that these compounds retarded the shoot elongation of tomato plants and the lengthening of leaves, roots and coleoptiles of wheat seedlings. Nuarimol was the most active compound, an important decrease in the C-4-desmethyl sterols of barley shoots was observed. Growth retarding effects of nuarimol after a seed dressing treatment of cereals at recommended rates were mentioned by several authors [193,194,195]. The effects were regarded as being transitory. In apple trees, Abdel-Rahman [196] reported reduction of terminal shoot elongation and of leaf area. He noted a prevention of fruit thinning after season-long treatment with fenarimol (and triadimefon) at the concentration of 8 oz./100 gal.

Buchenauer and Röhner [51] found that the imidazole compound *imazalil* caused shortening of coleoptiles, primary leaves and roots in wheat and barley when applied as a seed dressing at 25 or 50 g a.i./100 kg seed. This dosage is high in comparision to that needed for the control of Pyrenophora-diseases in barley (2–4 g/100 kg). At this rate, the authors found a 40% reduction of C-4-desmethylsterols and an accumulation of C-4-methylsterols and C-4,4-dimethylsterols.

In field experiments in Canada, imazalil caused a significant shortening of subcrown internodes in spring wheat at 20 and 30 g a.i./100 kg seed [197].

When formulated as an emulsifiable concentrate, the compatibility of *prochloraz* seems to be critical in repeated spray programs on dicotyledonous plants [198]. Since prochloraz can form stable complexes with manganese and copper, Birchmore et al. [198] reported improvement of crop safety, especially when powder formulations of the metal complexes were applied.

In contrast to the usual observation, that SBI-fungicides cause stronger phytotoxic symptoms in dicotyledonous than in monocotyledonous plants, Martin and Edgington [61] reported that *fenapanil* was more phytotoxic in barley than in soybeans after root application. Symptoms described in barley were stunting and chlorosis of the foliage with some necrosis developing at the leaf tips.

Whereas possible side effects of the weakly systemic triazole compound *fluotrimazole* have only incidently been reported [199], much work has been done on those of the related compounds *triadimefon* and *triadimenol*. Förster et al. [200,201,202] intensively studied the influence of both compounds on barley plants. Similar to effects caused by nuarimol, both fungicides caused retardation of shoot-, root- and coleoptile growth, and a disturbed geotropism of the seedlings. The effects were found during the first two weeks after seeding. After 30 days, no significant differences in the fresh and dry weight of shoots of treated and untreated plants, could be found.

Cytokinin-like effects could be clearly detected in green house studies, but were less significant in field experiments. For example, when based on fresh wheight there was a reduced chlorophyll content of primary leaves. When based on leaf area chlorophyll content was unchanged. Other effects were diminished xanthophyll-, carotin- and RNA-, but unchanged DNA- and protein content [201]. Further typical effects were the delayed senescence of treated leaves, a reduction in water consumption, and an increase in free amino acids [202].

Pennypacker et al.[203] examined the effect of triadimefon on the growth of annual bluegrass (Poa annua) in growth chambers under different environmental regimes. Under hot and wet conditions, the dry weight of treated plants decreased significantly but under other conditions tested (hot/dry; cold/dry) they did not. The authors found that in the field under hot and dry conditions, there was an improved survival of treated plants.

As with other fungicides that have the same mode of action as triadimefon and triadimenol, PGR-activities are more readily visible at lower concentrations on dicotyledons than on monocotyledons. On coffee seedlings, shortened internodes and reduced leaf size [204], on pine seedlings decreased root growth [205], on red raspberry earlier ripening and shorter and thinner canes [206], and on the apple variety 'Starking Delicious' an increase in fruit set [207] have been reported. Tueske and Kovacs [208] studied the therapeutic index of triadimefon in a model experiment with the host-parasite combination broad bean (Vicia faba)/Uromyces vicia-fabae.

Contrary to triadimefon and triadimenol, the locosystemic fungicide *bitertanol* hardly causes growth-regulator symptoms [42,100].

Buchenauer et al. [171] used barley seedlings to investigate the side effect of *propicon-*

azole and *etaconazole* after seed treatment with 25–50 g a.i./100 kg seed. In addition to a marked inhibition of lengthening of leaves, roots and coleoptiles (with etaconazole being the more active substance), the authors found a delay in chlorophyll destruction and an increased seedling tolerance to dryness, frost, and salt stress. In several reports on the performance of etaconazole on dicotyledonous plants, distinct growth-regulator effects were described. Paulus et al. [209] reported darker foliage and shortened internodes in roses and stunting effects on carnation. In apples, retarded tree growth, and smaller, thicker and darker green leaves in treated plants as compared to untreated were reported [58,42].

In the above mentioned study of Buchenauer et al. [171], *diclobutrazol* caused stronger PGR-effects than etaconazole and propiconazole. Bent and Skidmore [124] report a compaction of plant growth with shortening of internodes when diclobutrazole was applied as a spray to several plant species in green house tests. Soil drenches tended to have greater plant growth effects than sprays; dicotyledonous plants were more affected than monocotyledonous plants.

A final evaluation of the possible consequences of typical PGR-fungicide side effects is difficult since most reports have not been compared with yield measurements in plants without fungal infections.

As Tüske and Kovacs [208] mentioned, it can be presumed that side effects are sometimes favourable, in most cases indifferent and sometimes detrimental. The same authors added, that from their experience, 20–30% retardation of plant growth will still not cause significant differences in the yield of plants. As cited in connection with individual compounds, some changes have often been speculatively correlated with yield influencing factors like frost-, salt- and drought tolerance, fruit set, and retarded senescence of leaves. Although the application rates and techniques in such experiments are often not representative of those used in the field, this kind of observation should be the starting-point of accurate experiments in which the possible influence of side-effects on yield can be evaluated.

4.4 Morpholines

The morpholine fungicide *tridemorph* does not appear to inhibit gibberellic acid synthesis in plants. Benveniste and coworkers [210, 211] found that tridemorph blocked the cycloeucalenol-obtusifoliol isomerase in suspension cultures of Rubus fructiosus cells [210] and in whole maize plants [211]. This caused an accumulation of 9β,19-cyclopropyl-sterols and a dramatic loss of the usually predominating sterols stigmasterol and sitosterol.

Coinciding with the hypothesis that morpholine fungicides do not interfere with the biosynthesis of gibberellic acids in plants is the fact that typical growth-regulator effects have not been reported. Rather, unspecific signs of phytotoxicity such as leaf necroses have been observed.

Bergmann et al. [212] developed a physiological test system which uses the activity of oxidoreductases in plants as a measure of the phytotoxicity of tridemorph. This system showed that the sensitivity of the species decreased in the order cucumber > wheat > barley. The authors also noted variable responses of different winter wheat varieties to tridemorph as did Siddiqui [213] with some wild and cultivated grasses.

As with triforine, also tridemorph decreased the amino acid content of root exudates of wheat plants [181]. Thomas and Bingham [214] report abnormalities in esterase activation in Festuca pratensis. Swinburne [215], who investigated the phenylalanine-ammonia lyase (PAL) activity in barley seedlings found a reduction of the enzyme activity by tridemorph. Finally, Cortes et al. [216] reported that tridemorph interfered with the mitosis in meristematic cells of Allium cepa at higher concentrations.

Deymann [217] observed that wheat and barley seedlings tolerated a tridemorph treatment only when they were at least 14 days old. If younger plants were treated, they exhibited severe signs of phytotoxicity (necrotic spots on leaves) and died.

5 References

1. Fuchs, A., Davidse, L. C., de Waard, M. A., de Wit, P. J. G.: Pestic. Sci. *14*, 272–293 (1983)
2. Bayer AG, Marketing Research
3. Dekker, J.: In: Reference *4*, 177–186
4. Dekker, J., Georgopoulus, S. G. (Eds.): Fungicide resistance in crop protection. Center for Agricultural Publishing and Documentation, Wageningen (NL) (1982)
5. de Waard, M. A., Fuchs, A.: VI. Symposium Systemische Fungizide und antifungale Verbindungen. Reinhardtsbrunn 1980. pub. 1983, Akademie-Verlag Berlin, pp. 429–436
6. Symposium: Resistance to fungicides in plant pathogens: abstracts of papers, Neth. J. Pl. Path. 87, 233–255 (1981)
7. Drandarewski, C. A., Schicke, P.: Z. Pfl. Krankh. Pfl. Schutz *83*, 385–396 (1976)
8. Paul, V. H., Scheinpflug, H.: Pflanzenschutz-Nachrichten Bayer *32*, 83–92 (1979)
9. Buchenauer, H., Kemper, K.: Med. Fac. Landbouww. Rijksuni. Gent. *46*, 909–921 (1981)
10. Brandes, W., Paul, V.: Pflanzenschutz-Nachrichten Bayer *34*, 48—59 (1981)
11. Kuck, K. H., Scheinpflug, H., Tiburzy, R., Reisener, H. J.: Pflanzenschutz-Nachrichten Bayer *35*, 209–228 (1982)
12. Schlüter, K., Weltzien, H. C.: Nachrichtenblatt Deut. Pflanzenschutz. Braunschweig *29*, 17–20 (1977)
13. Smolka, S., Wolf, G.: Pflanzenschutz-Nachrichten Bayer *36*, 97–128 (1983)
14. Paul, V. H., Brandes, W.: Pflanzenschutz-Nachrichten Bayer *36*, 21–37 (1983)
15. Hendrix, J. W.: Ann. Rev. Phytopathol. *8*, 111 (1970)
16. Warner, S. A., Sovocool, G. W., Domnas, A. J.: Phytochemistry *21*, 2135–2136 (1981)
17. de Waard, M. A., Ragsdale, N. N.: 5. Intern. Symposium Systemfungizide, Reinhardsbrunn 1977, Proceedings 187–194, Akademie-Verlag (1979)
18. Kraus, P.: Pflanzenschutz-Nachrichten Bayer *32*, 17–30 (1979)
19. Adlung, K. G., Drandarevski, C. A.: Proc. 6th Brit. Insectic. Fungic. Conf. 2, 577–586 (1971)
20. Drandarewski, C. A.: Z. Pfl. Krankh. Pfl. Schutz *79*, 582–594 (1972)
21. Fuchs, A., Drandarewski, C. A.: Z. Pfl. Krankh. Pfl. Schutz *80*, 403 (1973)
22. Anonymous: Technische Information Saprol (R), Celamerck
23. Rohrbach, K. U.: Mitt. Biol. Bundesanst. Land-Forstwirtsch., Berlin–Dahlem *178*, 147–148 (1977)
24. Heaton, J. B.: Queensl. J. Agric. Anim. Sci. *37*, 155–160 (1981)
25. Koffmann, W., Kable, P. F.: Plant Disease Reporter *59*, 586–590 (1975)
26. Arenstein, Z., Ben Gal, A., Zilberstein, Y., Gilpatrick, J. D.: Phytoparasitica *3*, 79 (1975)
27. Ramsdell, D. C., Nelson, J. W., Myers, R. L.: Phytopathology *65*, 229–232 (1975)
28. Byther, R. S., Bristow, P. R.: Proc. Am. Phytopathol. Soc. *4*, 202 (1977)
29. Lockhart, C. L.: Canada Agriculture *27*, 39–41 (1982)
30. Giroto, R.: Idia (No. 313/414) p. 29–38 (1974, publ. 1976)
31. Blancard, D., Jamme, M. de la, Brun, J. M.: Revue Horticole *235*, 23–29 (1983)
32. Fortugno, C.: Idia (No. 317/320) p. 26–33 (1974, publ. 1977)
33. Kreidl, H.: Obstbau Weinbau *12*, 344–345 (1975)
34. Mantinger, H., Vigl, J.: Obstbau Weinbau *12*, 341–344 (1975)

35. Ost, W.: Symposium Eigenschaften und Wirkungsverhalten von Triforin, Heidesheim 1976, Abstracts 6–8 (1976)
36. Kato, T., Tanaka, S., Ueda, M., Kawase, Y.: Agr. Biol. Chem. *38*, 2377–2384 (1974)
37. Kato, T., Tanaka, S., Yamamoto, S., Kawase, Y., Ueda, M.: Ann. Phytopath. Soc. Japan *41*, 1–8 (1975)
38. Anonymous, Data Sheet, "Denmert®", Sumitomo Chemical Comp.
39. Anonymous: Commercial Grower, March *24*, 1972, p. 463
40. Buchenauer, H.: Z. Pfl. Krankh. Pfl. Schutz *86*, 341–354 (1979)
41. Beraud, J. M., Gueguen, F., Leca, J. L., Tussac, M., Bouquet, G.: Def. Veg. *34*, 17–24 (1980)
42. Kelley, R. D., Jones, A. L.: Phytopathology *71*, 737–742 (1981)
43. Verheyden, C.: Med. Fac. Landbouww. Rijksuniv. Gent *46/3*, 955–960 (1981)
44. Roussel, C.: Phytiatr.-Phytopharm. *30*, 165–171 (1981)
45. Gueguen, F., Leca, J. L., Prove, P.: Med. Fac. Landbouww. Rijksuniv. Gent *46/3*, 939–948 (1981)
46. Anonymous: Technische Information "Rubigan®", Elanco
47. Buchenauer, H.: Proc. 1977 Brit. Crop Prot. Conf. Brighton *3*, 699–711 (1977)
48. Anonymous: Technical Information "Trimidal®", Elanco, Homburg, Germany
49. Casanova, A., Doehler, R., Farrant, D. M., Rathmell, W. G.: Proc. 1977 Brit. Crop Prot. Conf., 1–7
50. Doehler, R., Mertz, M. V.: Mitt. Biol. Bundesanst. Land-Forstwirtschaft, Berlin–Dahlem, *191*, p. 185 (1979)
51. Buchenauer, H., Röhner, E.: V. Intern. Symposium Systemfungizide, Reinhardsbrunn 1977, Akademie-Verlag, 175–185 (1979)
52. Anonymous: Technical Information Sheet "Fungaflor®, Fungazil®, Fecundal®", Janssen Pharmazeutica, 1983
53. Cayley, G. R., Hide, G. A., Tillotson, Y.: Pestic. Science *12*, 103–109 (1981)
54. Melin, P., Plaud, G., Tezenas du Montcel, H., Laville, E.: Fruits *30*, p. 301–306 (1975)
55. Birchmore, R. J., Brookes, R. F.: Copping, L. G., Wells, W. H.: Proc. 1977 Brit. Crop Prot. Conf., Vol. *2*, 593–598 (1977)
56. Schneider, A., Drosihn, G., Muench, F.: Gesunde Pflanzen *35* (4), 101–105 (1983)
57. de Saint-Blanquat, A., My, J.: La Defense des Vegeteaux *221*, 121–141 (1983)
58. Szkolnik, M.: Plant Disease *65*, 981–985 (1981)
59. Shabi, E., Elisha, E., Zelig, Y.: Plant Disease *65*, 992–994 (1981)
60. Dijkhuizen, J. P., Ogawa, J. M., Manji, B. T.: Plant Disease *67*, 407–409 (1983)
61. Martin, R. A., Edgington, L. V.: Pestic. Biochem. Physiol. *17*, 1–9 (1982)
62. Yoder, K. S., Hickey, K. D.: Plant Disease *65*, 998–1001 (1981)
63. Hoffmann, J. A., Waldher, J. T.: Plant Disease *65*, 256–259 (1981)
64. Hansen, K. E.: Tidesskrift for Planteavl. *85*, 77–92 (1981)
65. Grewe, F., Büchel, K. H.: Mitt. Biol. Bundesanstalt Land-Forstwirtsch., Berlin–Dahlem *151*, 208–209 (1973)
66. Kaspers, H., Grewe, F., Brandes, W., Scheinpflug, H., Büchel, K. H.: VIII. Intern. Plant Protection Congress Moscow, 398–401 (1975)
67. Leroux, P., Gredt, M., Fritz, R.: Phytiatrie-Phytopharmacie *25*, 317–334 (1976)
68. Kolbe, W.: Pflanzenschutz-Nachrichten Bayer *35*, 36–71 (1982)
69. Becker, G.: Pflanzenschutz-Nachrichten Bayer *31*, 145–150 (1978)
70. Siebert, R.: Pflanzenschutz-Nachrichten Bayer *29*, 303–309 (1976)
71. Rowell, J. B.: Plant Disease Reptr. *65*, 235–236 (1981)
72. Stritzky, W. v., Trägner-Born, J.: Med. Fac. Landbouww. Rijkuniv. Gent *43*, 911–915 (1978)
73. Meeus, P., Fraselle, J., Gepts, P., Froidment, F.: Parasitica *34*, 35–48 (1978)
74. Mielke, H.: Nachrichtenbl. Deut. Pflanzenschutzd. *34*, 129–132 (1982)
75. Frohberger, P. E.: Mitt. Biol. Bundesanst. Land-Forstwirtsch., Berlin–Dahlem *191*, 169–170 (1979)
76. Mielke, H.: Gesunde Pflanzen *33*, 257–267 (1981)
77. Kolbe, W.: Erwerbsobstbau *22*, 29–32 (1980)
78. Mantinger, H., Vigl, J.: Erwerbsobstbau *20*, 160–165 (1978)
79. Doerntlein, D., Rota, F.: Pflanzenschutznachrichten Bayer *31*, 181–193 (1978)
80. Kaspers, H.: Der deutsche Weinbau *11*, 452 (1979)
81. Wybou, A., Stripecke, W.: Pflanzenschutz-Nachrichten Bayer *33*, 123–151 (1980)

82. Kaspers, H., Patel, N. K.: Pflanzenschutz-Nachrichten Bayer *33*, 152–164 (1980)
83. Bechet, G. R.: South African Sugar Journal *62*, 85–88 (1978)
84. Frossard, P.: Fruits *33*, 91–99 (1978)
85. Kolbe, W.: Pflanzenschutz-Nachrichten Bayer *34*, 80–84 (1981)
86. Clark, T., Clifford, D. R., Deas, A. H. B., Gendle, P., Watkins, D. A. M.: Pestic. Sci. *9*, 497–506 (1978)
87. Buchenauer, H., Grossmann, F.: Z. Pflanzenkrankh. Pflanzenschutz *89*, 309–324 (1982)
88. Kuck, K. H.: Pflanzenschutz-Nachrichten Bayer, in preparation
89. Gasztonyi, M., Josepovits, G.: Acta Phytopat. Acad. Sci. Hun *13*, 403–415 (1978)
90. Gasztonyi, M., Josepovits, G.: Pestic. Sci. *10*, 57–65 (1979)
91. Krämer, W., Büchel, K. H., Draber, W.: Pestic. Chem. Human Welfare Environ. Proc. Intern. Congr. Pestic. Chem. 5th 1982 (pub. 1983), Vol *1*, 309–314 Eds.: Miyamoto, Y., Kearny, P. C., Pergamon Press, Oxford, UK
92. Frohberger, P. E.: Mitt. Biol. Bundesanst. Land-Forstwirtsch., Berlin–Dahlem *178*, 150–151 (1977)
93. Frohberger, P. E.: Pflanzenschutz-Nachrichten Bayer *31*, 11–24 (1978)
94. Trägner-Born, J., van den Boom, T.: Pflanzenschutz-Nachrichten Bayer *31*, 25–37 (1978)
95. Gasztonyi, M.: VI. Intern. Symposium Systemische Fungizide und antifungale Verbindungen. Reinhardsbrunn 1980. pub. 1983 Akademie-Verlag Berlin, p. 113–121
96. Deas, A. H. B., Clifford, D. R.: Pestic. Biochem. Physiol. *17*, 120–133 (1982)
97. Buchenauer, H., Röhner, E.: Z. Pfl. Krankh. Pfl. Schutz *89*, 385–398 (1982)
98. Deas, A. H. B., Clark, T., Carter, G. A.: Pestic. Sci. *15*, 63–70 (1984)
99. Deas, A. H. B., Clark, T., Carter, G. A.: Pestic. Sci. *15*, 71–77 (1984)
100. Brandes, W., Kaspers, H., Krämer, W.: 3rd Intern. Congr. of Plant Pathology, München, 1978, pub. in: Pflanzenschutz-Nachrichten Bayer *32*, 1–16 (1979)
101. Kraus, P.: Pflanzenschutz-Nachrichten Bayer *32*, 17–30 (1979)
102. Brandes, W., Kaspers, H., Krämer, W.: Pflanzenschutz-Nachrichten Bayer *32*, 1–16 (1979)
103. Anonymous: Technical Information Baycor, Bayer AG (1981)
104. Scheinpflug, H., van den Boom, T.: Pflanzenschutz-Nachrichten Bayer *34*, 8–28 (1981)
105. Birch, P. A., Rose, P. W., Wainwaight, A.: Proc. 1981, Brit. Crop Prot. Conf., 545–588 (1981)
106. Nieuwoudt, C. J. L.: Pflanzenschutz-Nachrichten Bayer *37*, 19–31 (1984)
107. Trägner-Born, J., Kaspers, H.: Pflanzenschutz-Nachrichten Bayer *34*, 1–7 (1981)
108. Büchel, K. H.: Burdick and Jackson Symposium; 188th ACS National Meeting, Philadelphia (1984)
109. Urech, P. A., Schwinn, F. J., Speich, J., Staub, T.: Proc. 1979 Brit. Crop Prot. Conf., 508–515 (1979)
110. Urech, P. A., Speich, J.: Phytiatrie-Phytopharmacie *30*, 21–26 (1981)
111. Smith, J. M., Speich, J.: Proc. 1981 Brit. Crop Prot. Conf., 291–297 (1981)
112. Anonymous: Technical Release, Ciba Geigy Corp., 1980
113. Latham, A. J.: Plant Disease *67*, 1136–1139 (1983)
114. Whitney, G.: Plant Disease *67*, 587 (1983)
115. Allen, R. N., Peasley, D. L.: Australasian Plant Pathology *10*, 24–25 (1981)
116. Mourichon, X., Beugnon, M.: Fruits *37*, 595–597 (1982)
117. Staub, Th., Schwinn, F., Urech, P.: IX. Intern. Congr. of Plant Protection, Washington, Abstracts of Papers, *309* (1979)
118. Szkolnik, M.: IX. Intern. Congr. of Plant Protection, Washington, Abstracts of Papers, *315* (1979)
119. Gilpatrick, J. D.: IX. Intern. Congr. of Plant Protection, Washington, Abstracts of Papers, *315* (1979)
120. Gutter, Y.: Z. Pfl. Krankh. Pfl. Schutz *89*, 332–336 (1982)
121. Schachnai, A.: Plant Disease *66*, 733–735 (1982)
122. Vogel, C., Staub, T., Rist, G., Sturm, E.: Pesticide Chemistry: Human Welfare and the Environment, Ed. J. Miyamoto & P. C. Kearny, Volume *1*, p. 303–308 Pergamon Press 1983
123. Eberle, J., Ruess, W., Urech, P. A.: Proc. 10th. Intern. Congress of Plant Prot., Vol. *1*, 376–383, Brighton 1983
124. Bent, K. J., Skidmore, A. M.: Proc. 1979 Brit. Crop Prot. Conf., 477–484 (1979)
125. v. Zitzewitz, W., Heckele, K. H.: Mitt. Biol. Bundesanst. Land-Forstwirtsch. *191*, p. 184 (1979)

126. Javed, Z. U. R.: Kenya Coffee, Vol. *46*, 239 (1981)
127. Baldwin, B. C., Higgins, T. E.: 6. Symposium Systemische Fungizide und antifungale Verbindungen, Reinhardsbrunn, Abstracts, *4* (1983)
128. Skidmore, A. M., French, P. N., Rathmell, W. G.: Proc. 10. Intern. Congr. of Plant Prot., Brighton 1983, 368–375
129. Northwood, P. J., Horellou, A., Heckele, K. H.: Proc. 10. Intern. Congr. of Plant Prot., Brighton 1983, p. 930
130. Funaki, Y., Ishiguri, Y., Kato, T., Tanaka, S.: Pestic. Chem. Human Welfare Environ. Proc. Intern. Congr. Pestic. Chem. 5th., 1982 (pub. 1983) Vol. *1*, 309–314
131. Pommer, E. H., Zeeh, B.: 4th. Intern. Congr. of Plant Pathology, Melbourne, Australien 1983
132. Shepard, M. C., French, P. N., Rathwell, W. G.: 10th. Intern. Congr. of Plant Prot. Brighton 1983, Proc. Vol. *2*, 521
133. Fort, T. M., Moberg, W. K.: Proc. 1984 Brit. Crop Prot. Conf., Vol. *2*, p. 413–419
134. Marchand, D., Serra, G.: La Defense Vegetable *27*, 144–152 (1973)
135. Watkins, J. E.: Plant Disease, *351* (1983)
136. Pommer, E. H., Himmele, W.: Intern. Symposium on Crop Prot., Gent, Belgium, Abstracts *51* (1979)
137. Hess, C., Heimes, R.: BASF Agricultural Bulletin *6*, 1–31 (1975)
138. Lehmann, Beyme: Pflanzenschutz- und Schädlingsbekämpfungsmittel (PSM) *3/1*, p. 20 (1983)
139. Bohnen, K., Pfiffner, A.: Med. Fac. Landbouww. Rijksuni. Gent *44*, 487–497 (1979)
140. Anonymous: Technical Information "Corbel®", BASF AG/Dr. R. Maag AG, 1982
141. Pommer, E. H.: Pestic. Sci. *15*, 285–295 (1984)
142. Perkow, W.: Wirksubstanzen der Pflanzenschutz- und Schädlingsbekämpfungsmittel, 2nd Ed. 1983, Verlag Paul Parey, Berlin and Hamburg
143. Edgington, L. V.: Annu. Rev. Phytopathol. *19*, 107–124 (1981)
144. Fuchs, A., de Vries, F. W., Aalbers, A. M. J.: Pestic. Sci. *7*, 115–126 (1976)
145. Fuchs, A., de Vries, F. W., Aalbers, A. M. J.: Pestic. Sci. *7*, 127–134 (1976)
146. v. Bruchhausen, V., Stiasni, M.: Pestic. Sci. *4*, 767–773 (1973)
147. Ohkawa, H., Shibaike, R., Okihara, Y., Morikawa, M., Miyamoto, J.: Agr. Biol. Chem. *40*, 943–951 (1976)
148. Brown, I. F., Hall, H. R.: Proc. 1981 Brit. Crop Prot. Conf., 573–578 (1981)
149. Cardoso, J. E., Hildebrandt, A. C., Grau, C. R.: Fitopatol. Bras. *4*, 11–15 (1979)
150. Vonk, J. W., Dekhuizen, J. N.: Med. Fac. Landbouww. Rijksuni. Gent *44*, 927–934 (1979)
151. Reisdorf, K., Wurzer-Fassnacht, U., Walther, H. F.: Z. Pfl. Krankh. Pfl. Schutz *90*, 641–649 (1983)
152. Cooke, B. K., Pappas, A. C., Jordan, V. W. L., Western, N. M.: Pesticide Science *10*, p. 467–472 (1979)
153. Martin, R. A., Edgington, L. V.: Pestic. Biochem. Physiol. *16*, 87–96 (1982)
154. Kelly, R. D., Jones, A. L.: Can. J. Plant Pathol. *4*, 243–246 (1982)
155. Buchenauer, H.: Mitt. Biol. Bundesanst. Land-Forstwirtsch., Berlin–Dahlem *165*, 154–155 (1975)
156. Buchenauer, H.: Pflanzenschutz-Nachrichten Bayer *29*, 267–280 (1976)
157. Kluge, E., Lyr, H.: Symposium Systemische Fungizide und antifungale Verbindungen, Reinhardsbrunn 1980, pub. 1982, Berlin 1 N 1983, 379–383
158. Sanders, P. L., Burpee, L. L., Cole, H., Duich, J. M.: Phytopathology *68*, 1482–1487 (1978)
159. Scheinpflug, H., Paul, V., Kraus, P.: Pflanzenschutz-Nachrichten Bayer *31*, 101–115 (1978)
160. Shillingford, C. A., Sinclair, J. B.: Plant Disease Reporter *62*, 1107–1111 (1978)
161. Paul, V.: Phytomedizin *9*, 18 (1979)
162. Führ, F., Paul, V., Steffens, W., Scheinpflug, H.: Pflanzenschutz-Nachrichten Bayer *31*, 116–131 (1978)
163. Brandes, W., Steffens, W., Führ, F., Scheinpflug, H.: Pflanzenschutz-Nachrichten Bayer *31*, 132–144 (1978)
164. Kraus, P.: Pflanzenschutz-Nachrichten Bayer *34*, 197–212 (1981)
165. Scheinpflug, H., Paul, V.: Neth. J. Pl. Path. *83*, 105–111 (1977)
166. Schlüter, K.: Z. Pfl. Krankh. Pfl. Schutz *84*, 612–614 (1977)
167. Jenkyn, J. E., Dyke, G. V., Todd, A. D.: Plant Pathology *32*, 311–324 (1983)

168. Steffens, W., Führ, F., Kraus, P., Scheinpflug, H.: Pflanzenschutz-Nachrichten Bayer *35*, 171–188 (1982)
169. Rathmell, G., Skidmore, M.: Outlook on Agriculture *11*, 37–43 (1982)
170. Szkolnik, M.: Plant Disease *67*, 360–366 (1983)
171. Buchenauer, H., Kohts, T., Roos, H.: Mitt. Biol. Bundesanst. Land-Forstwirtsch. Berlin-Dahlem, *203*, 310–311 (1981)
172. Otto, S., Pommer, E. H.: Med. Fac. Landbouww. Rijksuni. Gent *38*, 1493–1506 (1973)
173. Hamm, R.: BASF-Symposium Central America 1983, 15–24 (1983)
174. Sharma, S. R., Sohi, H. S.: Pestic. *14*, 21–24 (1980)
175. Vyas, S. C., Sinha, O. K., Joshi, L. K.: Pestic. *10*, 32–33 (1976)
176. Balasubramanyan, S.: Symposium Ergosterol Biosynthesis Inhibitors Reading, UK, *333* (1983)
177. Anonymous: Deutscher Gartenbau *31*, 50 (1977)
178. Attabhanya, A.; Holcomb, G. E.: Plant Disease Reporter *60*, 56–59 (1976)
179. Boudier, B.: Revue Forestiere Francaise *33*, 394–399 (1981)
180. Decallone, J., Soliman, M. S., Weyns, J., Meyer, J.: Med. Fac. Landbouww. Rijksuni. Gent *44*, 943–950 (1979)
181. Jalali, B. L., Domsch, K. H.: Phytopathol. Z. *90*, 22–26 (1977)
182. Church, R., Williams, R. R.: J. of Horticultural Science *53*, 91–94 (1978)
183. Bristow, P. R.: Plant Disease *65*, 350–353 (1981)
184. Globerson, D., Eliasi, R.: Euphitica *28*, 115–118 (1979)
185. Smith, J. W. M.: Euphitica *28*, 351–359 (1979)
186. Shive, J. B. jr., Sisler, H. D.: Plant Physiol. *57*, 640–644 (1976)
187. Coolbaugh, R. C., Hamilton, R.: Plant Physiol. *57*, 245–248 (1976)
188. Sisler, H. D., Ragsdale, N. N., Waterfield, W. F.: Symposium Ergosterol Inhibiting Fungicides, Reading, GB (1983)
189. Goldsmith, I. R., Hood, K. A., MacMillan, J.: Symposium Ergosterol Inhibiting Fungicide, Reading, GB (1983)
190. Schmitt, P., Benveniste, P.: Phytochemistry *18*, 1659–1665 (1979)
191. Buchenauer, H., Roehner, E.: Pestic. Biochem. Physiol. *15*, 58–70 (1981)
192. Buchenauer, H.: Ber. Deutsch. Bot. Ges. *96*, 427–457 (1983)
193. Casonova, A., Doehler, R., Farrant, D. M., Rathmell, W. G.: Proc. 1977 Brit. Crop Prot. Conf., 1–7
194. Doehler, R., Mertz, M. V.: Mitt. Biol. Bundesanst. Land-Forstwirtsch., Berlin-Dahlem, *191*, p.185 (1979)
195. Luz, W. C., Vicara, J. C.: Plant Disease *66*, 135–136 (1982)
196. Abdel-Rahman, M.: Proc. Anm. Phytopathol. Soc. *4*, 213 (1977)
197. Chinn, S. H. F., Verma, P. R., Spurr, D. T.: Can. J. Plant Sci. *60*, 1467–1472 (1980)
198. Birchmore, R. J., Wells, W. H., Copping, L. G.: Proc. 1977 Brit. Crop Prot. Conf., 583–588 (1979)
199. Byford, W. J.: Anm. Appl. Biol. *86*, 47–57 (1977)
200. Foerster, H., Buchenauer, H., Grossmann, F.: Z. Pfl. Krankh. Pfl. Schutz *87*, 473–492 (1980)
201. Foerster, H., Buchenauer, H.. Grossmann, F.: Z. Pfl. Krankh. Pfl. Schutz *87*, 640–653 (1980)
202. Foerster, H., Buchenauer, H., Grossmann, F.: Z. Pfl. Krankh. Pfl. Schutz *87*, 717–730 (1980)
203. Pennypacker, B. W., Sanders, P. L., Gregory, L. V., Gilbride, E. P., Cole, H., jr.: Can. J. Plant Pathol. *4*, 259–262 (1982)
204. Venkatasubbaiah, P., Muthappa, B. N.: J. Coffee Res. *11*, 1–3 (1981)
205. South, D. B., Kelly, W. D.: Can. J. For. Res. *12*, 29–35 (1982)
206. Pepin, H. S., Macpherson, E. A., Clements, S. J.: Can. J. Plant Scie. *60*, 1203–1208 (1980)
207. Strydom, D. K., Honeyborne, G. E.: Hort Science *16*, 51 (1981)
208. Tueske, M., Kovacs, M.: Z. Pfl. Krankh. Pfl. Schutz *90*, 460–467 (1983)
209. Paulus, A. O., Nelson, J., Besemer, S., Brendler, S., Hall, B.: Med. Fac. Landbouww. Rijksuni. Gent *45* (2), 245–252 (1980)
210. Schmitt, P., Benveniste, P., Leroux, P.: Phytochemistry *20*, 2153–2158 (1981)
211. Bladocha, M., Benveniste, P., Leroux, P.: Plant Physiol. *71*, 756–762 (1983)
212. Bergmann, H., Müller, R., Neuhaus, W.: VI. Intern. Symposium Systemische Fungizide und antifungale Verbindungen, Reinhardsbrunn 1980, Akademie-Verlag, Berlin, 355–359 (1983)
213. Siddiqui, K. A.: Experientia *28*, 1239–1240 (1972)

214. Thomas, H., Bingham, M. H.: Welsh Plant Breeding Station, Report for 1976, 131 (1977)
215. Swinburne, T. R.: Physiol. Plant Path. *5*, 81–88 (1975)
216. Cortes, F., Escalza, P., Moreno, J., Lopez-Campos, J. L.: Cytobios *34*, 181–190 (1982)
217. Deymann, A.: Dissertation, University of Bonn (1981)

Anti-Feeding Compounds

Dale M. Norris
Entomology and Graduate Neurosciences,
642 Russell Laboratories, University of Wisconsin,
Madison, WI 53706, USA

Chemical prevention or reduction of insect feeding is discussed. Presented aspects include molecular structures, behavioral assays, occurrences of inhibitory molecules in candidate host plants and animals sensory ultrastructure, neurophysiology, biochemistry and biophysics of chemical perception, encoding of chemical energy states into nerve impulses and insect avoidance behavior, practical usefulness of antifeedants in integrated pest management (IPM) programs and possible future developments. An unique comprehensive presentation of dynamic factual information is attempted. The reader should gain the best possible understanding of the subject, chemical antifeedants. Current knowledge of this subject strongly indicates that significant developments and practical uses of antifeedants in IPM programs should expand in the near future. Future investigations of insect perceptions of chemicals as antifeedants also should yield significant contributions to the basic understanding of neurobiology and especially information encoding in neurons.

1 Introduction

Chemical antifeedants for insects are substances which, when perceived, reduce or prevent feeding. Feeding is the ingestion of a substrate which may be a solid, gel, sol or liquid. Feeding here includes the act of "drinking" as traditionally applied to the ingestion of a liquid. A substrate is ingested to obtain nutrients, water or bulk. Substrates may be biotic or abiotic. If an insect feeds on a live substrate, then the latter frequently is termed a "host". If the live substrate is killed by the feeding process then it becomes "prey". If feeding does not occur, the live organism is a "non-host" with regard to ingestion by the insect.

Chemical antifeedants vary in volatility; some are sufficiently volatile to contact insects in habitats removed significantly from the site of chemical release. If these compounds not only prevent feeding on a candidate substrate, but also prevent, or reduce, contact between the insect and that substrate then such chemicals also may be termed repellents. Many compounds thus possess a spectrum of properties which may affect other insect behaviors besides feeding.

Because a chemical which is an antifeedant for one insect species or biotype may be an excitant to the feeding of other insects, discussions of an antifeedant should include the involved species or biotype.

2 Antifeedants in Chemical Ecology

The molecules, ions or free radicals which are operative as interspecific messengers in chemical ecology frequently are classified as allomones (adaptive to the emitter) or kairomones (adaptive to the perceiver) [1]. Each messenger actually is allomonic to some species or biotypes and kairomonic to others [2]; it thus is a "double agent" which tells some organisms that they are „in a favorable habitat" and others that they are "in hostile territory". This "yin-yang" type of principle in chemical ecology [3] provides a functional basis for determining the interspecific and intraspecific abundances among and within habitats and ecological niches, i.e., species and biotypes will be relatively abundant in situations dominated by their kairomones and relatively rare in places dominated by their allomones [2]. The larger reality which emerges from this understanding of chemical ecology is that the distribution of life on earth is very ordered (non-random).

Because antifeedants are the messengers in chemical ecology which reduce or prevent organismal ingestion, they determine which substrates are consumed by an insect. They further determine within a substrate (e.g., host plant) which specific portions may be eaten. The aglycone 2,4-dihydroxy-7-methoxy-2H-1,4-benzoxazin-3-one (DIMBOA) in young leaf tissues of some *Zea mays* plants

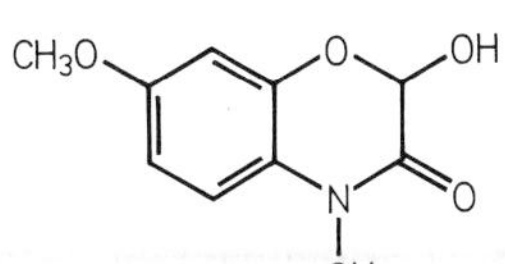

Dimboa

(1)

is a major antifeedant to first-instar larvae of the European corn borer, *Ostrinia nubilalis*[4]. If this compound and some related molecules are greatly reduced in concentration in such corn tissues, the borer larvae are strongly stimulated to feed by chemical excitants which are present in the plant. In such intact *Z. mays* plants, the borer larvae feed readily in older tissues where the concentration of these antifeedants is much less. Likewise, juvenile needles of the pine tree, *Pinus banksiana*, are not consumed by larvae of two species of *Neodiprion* sawflies because they contain antifeedant concentrations of 13-keto-8(14)-podocarpen-18-oic acid[5]. This compound is not

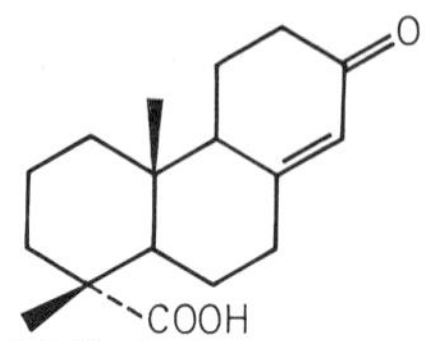

13-keto-8(14)-Podocarpen-18-oic acid

(2)

present in such concentrations in older needles, and the insect larvae eat such tissues. This is an excellent example of coevolution between an insect herbivore and a plant such that the insect only consumes those needles which are not essential to the plant's survival. Removal of the juvenile needles of *P. banksiana*, e.g., by a herbivore's feeding, commonly kills the tree.

3 Antifeedants in Non-Host Decisions

Present knowledge indicates that the presence of allomones (e.g., antifeedants), not the lack of kairomones (e.g., feeding excitants), more frequently provides the major chemical basis for an insect's decision that a living substrate (organism) is a non-host with regard to ingestion[6]. As an example, the non-host status of hickory (*Carya*) trees with regard to adult *Scolytus multistriatus* feeding was attributed especially to the aglycone 5-hydroxy-1,4-naphthoquinone (juglone), a potent allomone (antifeedant) to this insect[7, 8].

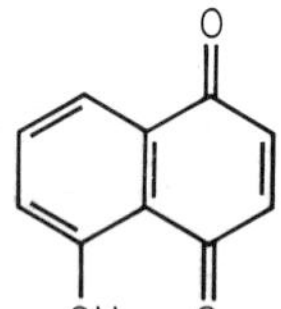

Juglone

(3)

When juglone was removed from the extractables from *Carya* tissues, then *S. multistriatus* fed readily on the remaining substrate. *Carya* thus contains feeding excitants for these beetles, but they naturally are masked by the dominant antifeedant juglone in the non-host plant tissues. Gossypol and

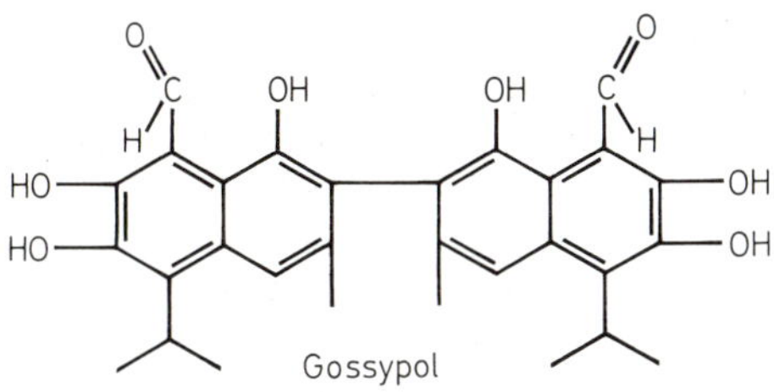

Gossypol (4)

related terpenoids in cotton (*Gossypium*) plants make them non-hosts to *Heliothis* and *Epicauta* [9,10,11]. The different non-host ranges within the Solanaceae for larvae of three species of insects (i.e., *Lema trilineata daturaphila*, *Leptinotarsa decemlineata* and *Manduca sexta*) further demonstrate a dominant role for antifeedants in non-host/host decisions by these insects [12]. These specific non-host ranges are determined by the differential susceptibilities of the three larvae to the various classes of alkaloids in the Solanaceae. *Lema* has the most restricted host range because both nicotine and most steroidal alkaloids are allomones (i.e., antifeedants) to its larvae. *Leptinotarsa* larvae have an intermediate host range because they can cope with nicotine and certain of the widely distributed steroidal alkaloids,

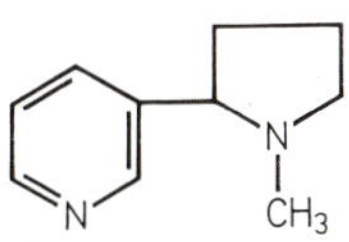

(5)

Nicotine

but not with tropane and other alkaloids. Finally, *Manduca* copes with all three classes of alkaloids in Solanaceae and thus has a relatively broader host range in the plant family. There is no evidence that these non-host/host decisions in solanaceous plants involve specific feeding excitants.

4 Experimental Evaluations

Because any one chemical is an antifeedant to only some insect species, experimental evaluations should involve a variety of species. For example, warburganal is one of the more potent antifeedants against the monophagous *Spodoptera exempta*, but is hardly allomonic to the tobacco hornworm, *Manduca sexta*, or the vagrant grasshopper, *Schistocerca vaga* [14]. The latter two species are oligophagous-polyphagous.

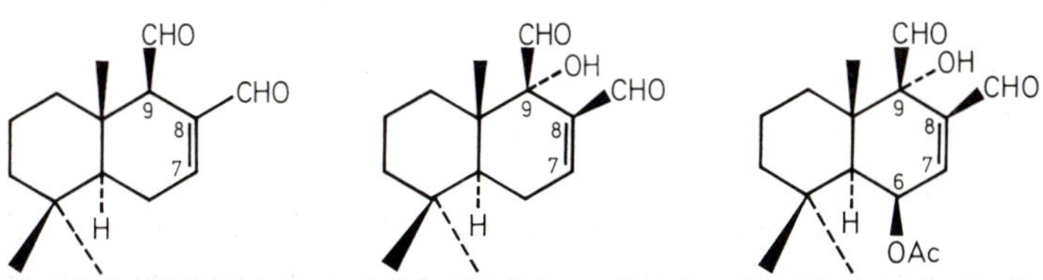

(6)

Polygodial Warburganal Ugandensidial

It seems that antifeedant activity by a chemical can be detected more efficiently through assays with monophagous (specialist) species rather than polyphagous (generalist) insects. This is logical because allomones (e.g., antifeedants) frequently determine which plant species are non-hosts for an insect, and a monophagous species has such a non-host relation with more plant species than a polyphagous insect does. Therefore, the specialist insect species or biotype should detect the greater variety of antifeedant chemicals.

Initial experimental evaluation of a candidate chemical should involve an insect bioassay in which the relative amount of feeding on treated versus positive (i.e., feeding excitatory)-control substrates is determined. Design of a bioassay should be based on a detailed analysis of the environmental parameters, other than the antifeedant, which affect the insect's feeding activities. The protocol and logistics of the bioassay thus should incorporate considerations of all critical parameters (e.g., light; temperature; substrate for presenting the treatment chemical; relative humidity; chemical stability; dimensions of the assay arena; concentration of treatment chemical; duration of assay; and stage, physiological condition and age of the insect).

A bioassay presenting a treatment chemical or fraction applied to a positive control versus the positive (excitatory) control alone in a common arena constitutes a "choice" test (Fig. 1). Such a choice is usually preferred over a "no-choice" test which presents only the candidate (treatment) chemical applied to the positive control. If insects are presented with only one substrate which naturally is a non-host, over time many species may feed on it; whereas, they would not feed on it in the free-choice, natural situation. Thus, such a no-choice (one substrate) test frequently proves to be too severe an assay. The assay substrate for a choice test frequently is a standard portion of the host, e.g., an uniform-diameter disk of leaf tissue, a whole leaf or a twig section from a tree branch. In such a choice test, at least two uniform samples of this positive-control substrate (e.g., leaf disk) are presented to the insect. One sample is treated

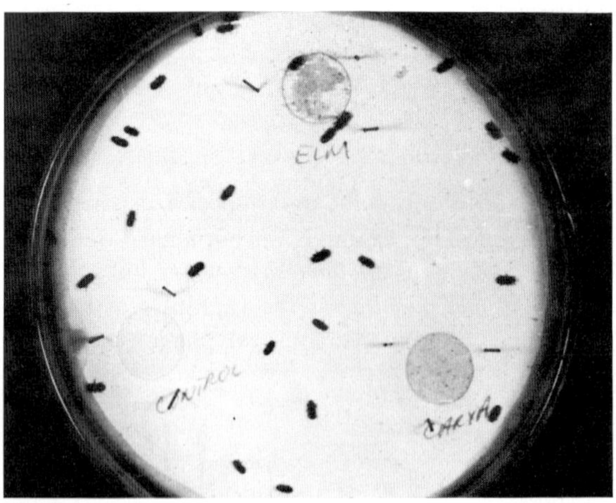

Fig. 1. A "choice" bioassay involving the adult smaller European elm bark beetle (*Scolytus multistriatus*) with a presented choice among elderberry-pith disks soaked in host elm (*Ulmus*) extract, nonhost *Carya* extract or solvent (chloroform). Feeding has occurred only on the disk treated with host elm extractables

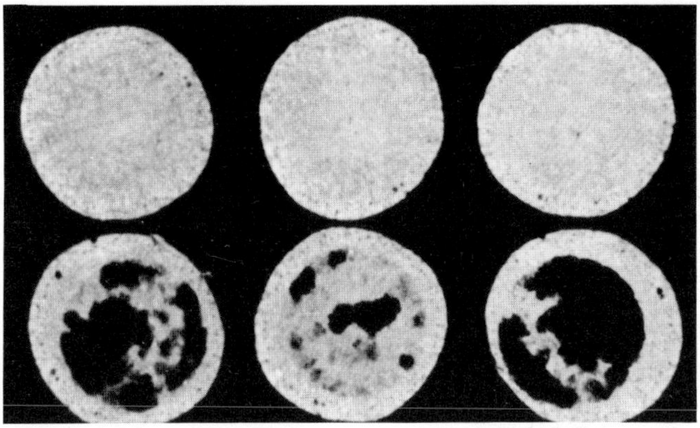

Fig. 2. Disks of a standardized "neutral" substrate, elderberry pith, upon which candidate pure chemicals or mixtures can be bioassayed against various chewing insects to determine their effects on the insect's feeding. Upper row of three disks was treated only with solvent before exposure to test insects; whereas, lower row was also treated with a feeding excitant

with the candidate chemical and the other is treated with only the solvent which was used to dissolve or dilute the candidate chemical. The latter substrate plus solvent thus is a positive control.

Many bioassays involve a standard "neutral" substrate. Such a substrate does not itself excite or inhibit the insect's feeding response (Fig. 2). Uniform disks of elderberry pith have been widely used as a neutral substrate. Disks of filter paper or powdered glass also may be so employed. Disks (or plugs) of agar-based media frequently are so used for insects that need to tunnel into the substrate as they feed. Plugs or pellets of pressed, powdered cellulose also function well as a neutral substrate for some insect species (e.g., *Periplaneta americana*) (Fig. 3).

Besides using feeding bioassays, antifeedants may be evaluated via electrophysiological (e.g., electroantennogram, EAG) techniques (Fig. 4). Such methods may be very sensitive, but the results initially need to be positively correlated with data from insect feeding bioassays. Electrophysiologically based assays are extremely valuable for identifying which sensilla on the insect are sensitive to antifeedants. These assays also are useful in studying the effects of concentration on antifeedant activity. A major disadvantage of electrophysiological techniques is the requirement for expensive, rather sophisticated equipment.

Antifeedant activity also may be detected through in vitro binding (e.g., receptor) studies (Fig. 5). Such chemicals naturally must variously interact (e.g., bind or complex) with macromolecular components of the dendritic membranes of the chemosensory neurons in appropriate sensilla. Thus, the in vitro exposure of a standardized aliquot of appropriate receptor macromolecules (e.g., lipoproteins) from such sensory membranes to candidate antifeedants versus positive controls, solvents, etc., under uniform conditions may detect and compare antifeedant activities. Quantitation of the activity is commonly based on bound radioactivity, changes in fluorescence or other spectroscopic measurements (e.g., UV or infrared) which are attributable to the presence of the treatment chemical. Results from such binding assays also need to be positively correlated with those from insect feeding bioassays.

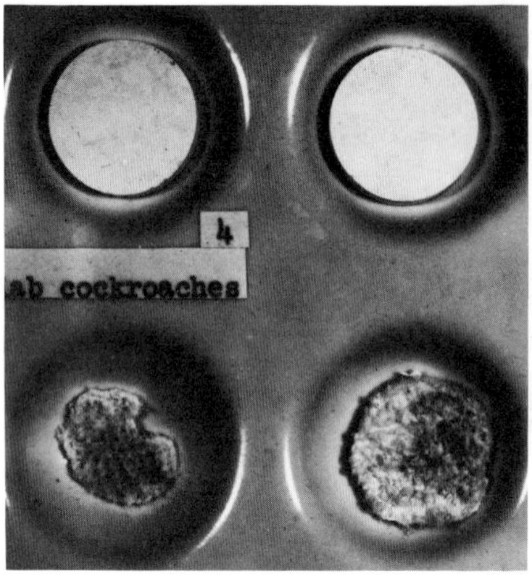

Fig. 3. Pellets of pressed, powdered cellulose used as a "neutral" substrate upon which candidate pure chemicals or mixtures can be bioassayed against chewing insects (e.g., *Periplaneta americana*, American cockroach) to determine effects on feeding. Upper two disks were only treated with solvent before exposure to cockroachs; whereas, lower two disks were also treated with a feeding excitant

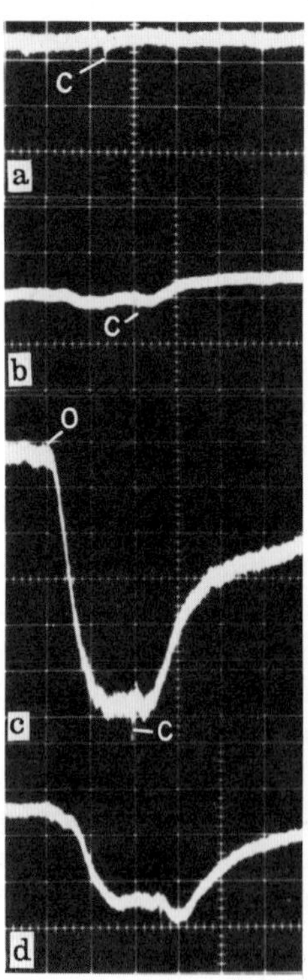

Fig. 4a–d. Electroantennograms of a male *P. americana* antenna elicited by the excitant amyl acetate **c**, and as inhibited (i.e., hyperpolarized) by a prior pulse of the antenna with the antifeedant 5-hydroxy-1, 4-naphthoquinone (juglone) **d**. Baseline is shown in **a**, and baseline plus the standard pulse of control (untreated) air **b**. (1 mV per grid unit)

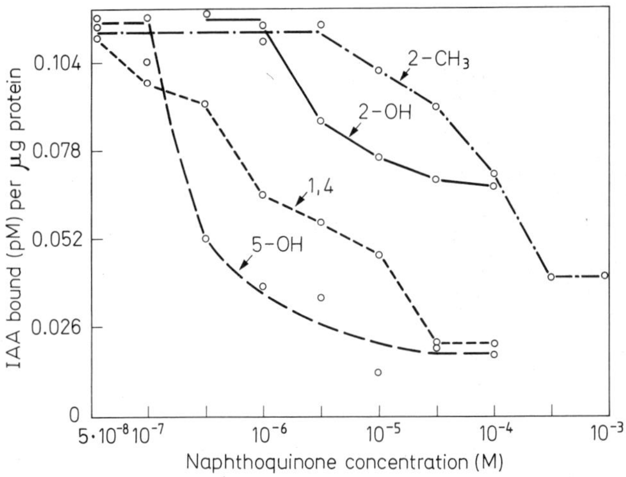

Fig. 5. Average picomoles of ^{14}C-labelled iodoacetic acid (IAA) subsequently bound in electrophoretically purified redox-complex (receptor) lipoprotein when 10^{-4} M IAA was challenged for receptor sites by a 15-minute preincubation of the lipoprotein with increasing concentrations of each of four messenger 1,4-naphthoquinone antifeedants

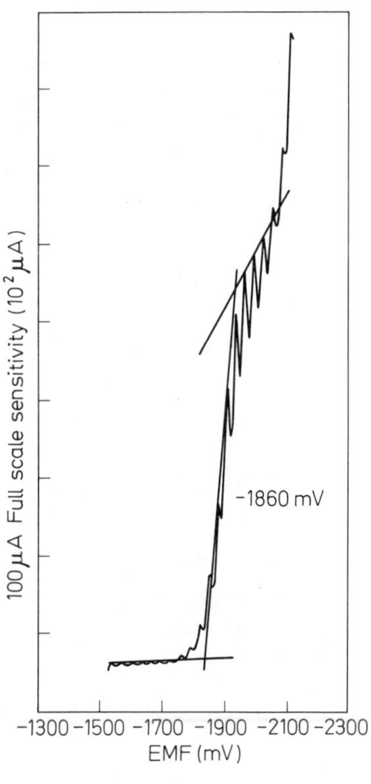

Fig. 6. Dropping-mercury electrode (DME) polarograph of Triton X-100 solubilized lipoprotein containing the redox-complex (receptor) protein. Test sample contained 100 μg protein per ml of potassium phosphate (0.1 M), pH 7.0 which contained 25% (v/v) of 95% ethanol. Sample was deoxygenated with dry nitrogen before and during each analysis. The characteristic E 1/2 value for the redox-complex protein was —1860 mV in the absence of antifeedant chemicals

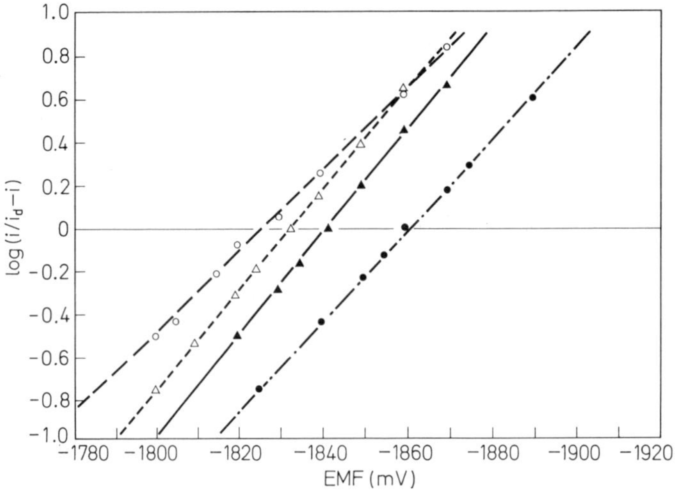

Fig. 7. Logarithmic analyses of the polarographic waves obtained from the Triton X-100 solubilized, redox-complex protein from chemosensory nerves in the antennae of the WARF biotype of adult *Periplaneta americana* males following the indicated treatment. Protein alone (control) (black circles), protein + 2-methyl-1,4-naphthoquinone (mendadione) (black triangles), protein + 1,4-naphthoquinones (open triangles) and protein + 5-hydroxy-1,4-naphthoquinone (juglone) (open circles). The applied electromotive force (EMF) at which the log i/i_d-i is zero gives the half-wave potential (E 1/2) for the protein under the specified treatment condition

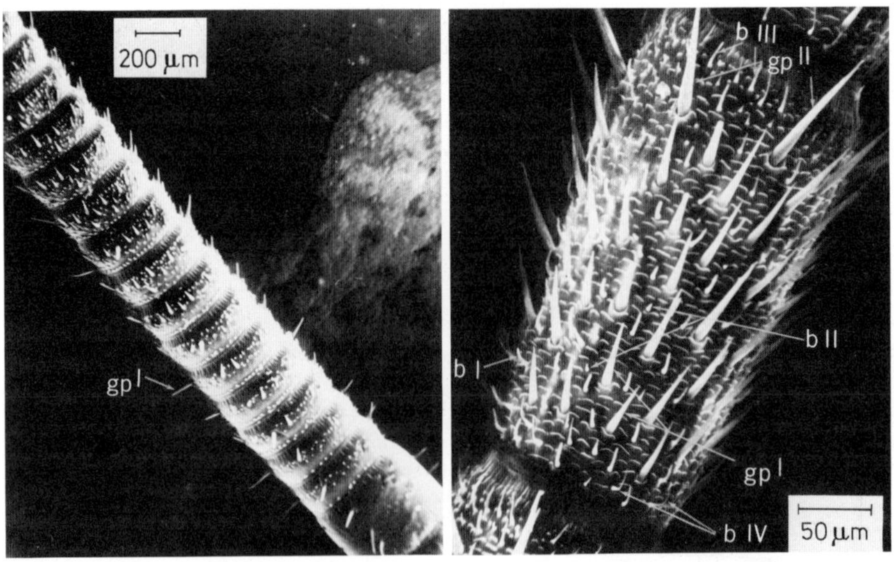

Fig. 8. Basal segments of antennal flagellum of adult *P. americana* male, left (scale = 200 µm); segment of mid-region of flagellum, right (scale = 50 µm). gpI, gpII, grooved peg, types I and II; bI, bII, bIII and bIV, types of sensilla basiconica

Another type of useful in vitro assay involves electrochemistry (e.g., dropping-mercury-electrode polarography, Fig. 6). For a chemical to inhibit or prevent insect feeding, it must enter into an energy-transfer interrelationship with the appropriate energy-transducing receptor macromolecule in the sensory neuron. In some instances this energy transfer may be detected via electrochemistry, and measured quantitatively in terms of the millivolt shift in the isolated receptor's half-wave potential (i.e., E 1/2 value, Fig. 7). Our laboratory has especially used this physicochemical technique to quantify the dynamic, reversible interactions of several antifeedant 1,4-naphthoquinones with sensory receptor macromolecules in *Periplaneta americana* (American cockroach) antennae (Fig. 8). Results from this in vitro technique also should be correlated with the findings from insect feeding bioassays.

5 Antifeedants in Plants

It is generally agreed that many phytophagous insects' rejections of plant species as food substrates are attributable significantly to the presence of allomones (e.g., antifeedants)[6, 12, 15]. Antifeedants in plants vary greatly in their chemistry and include inorganics, primary and intermediate metabolites and secondary compounds. Most of the molecules, ions or free radicals which are involved as messengers in chemical plant ecology apparently have allomonic (e.g., antifeedant) activity to one or more insect species or biotypes[2]. A list of compounds having antifeedant actions to one or more insect species therefore is likely to be limited mostly by our ignorance of the subject. Because of their diverse chemistries, most plants contain antifeedants for numerous phytophagous insect species. Within the theoretically available chemical armoury, different families of plants frequently rely on one to a few major compounds for antifeedant activities[16]. The various compartmentalizations of such chemicals in plants also may greatly influence their acessibilities, and thus their allomonic actions to specific insects[17].

5.1 Secondary Compounds

The greater diversity of antifeedants is found in the advanced versus primitive plants (e.g., angiosperms versus bacteria), and this is attributed to a greater array of secondary metabolites (i.e., secondary compounds)[16]. These latter chemicals have long been considered as major parameters used by insects in their rejection of specific plants as food.

Several classes of secondary plant compounds are listed in Table 1. Within this diverse group, alkaloids, flavonoids and terpenes apparently dominant numerically

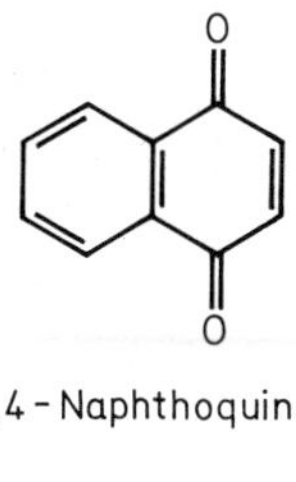

(7)

1,4-Naphthoquinone

as antifeedants to one or more insects. Other smaller classes, e.g. 1,4-benzoquinones and 1,4-naphthoquinones, are also extremely important because of their potent antifeedant effects on broad arrays of insect species.

The diversity of compounds within a secondary chemical class in higher plants is well illustrated by the terpenes (Fig. 9). They arise from the condensation of two or

Table 1. Secondary products which have allomonic (e.g., antifeedant) activity against one or more species of insects[a]

Chemical class	Estimated number of known structures	Example
Acetylenes	750	Dihydromatricaria
Alkaloids	4,500	Nicotine
Amino acids	250	Canavanine
Carotenoids	300	Fucoxanthin
Coumarins	150	Coumarin
Cyanogenic glycosides	50	p-Hydroxymandelonitrile glucoside
Flavonoids	1,200	Quercetin
Glucosinolates	80	Sinigrin
Lignins	50	Excelsin
Phenolic acids	100	p-Hydroxybenzoic acid
Quinones	200	Juglone
Terpenes	1,100	Glaucolide-A
Steroids	600	Ecdysteroids
Proteins	?	Lectins

[a] Adapted largely from [16]

Table 2. Representative alkaloids which possess antifeedant activity against one or more insect species

Chemical	Insect	Ref.
Berberine	*Leptinotarsa decemlineata*	24)
Cocaine	*L. decemlineata*	24)
Chaconine	*L. decemlineata*	18)
Chinchonine	*L. decemlineata*	24)
Demissine	*L. decemlineata*	18)
Isoboldine	*Abraxas miranda*	29)
Leptine	*Leptinotarsa decemlineata*	18)
Leptinine I	*L. decemlineata*	18)
Leptinine II	*L. decemlineata*	18)
Nicandrenone	*L. decemlineata*	12)
Nicotine	*L. decemlineata*	12)
Papaverine	*L. decemlineata*	24)
Sanguinarine	*L. decemlineata*	24)
Solanine	*L. decemlineata*	18)
Strychnine	*L. decemlineata*	24)
Tomatine	*L. decemlineata*	18, 24)
	Empoasca fabae	20)
	Manduca sexta	21)
	Dysdercus spp.	22)
	Pieris brassicae	23)

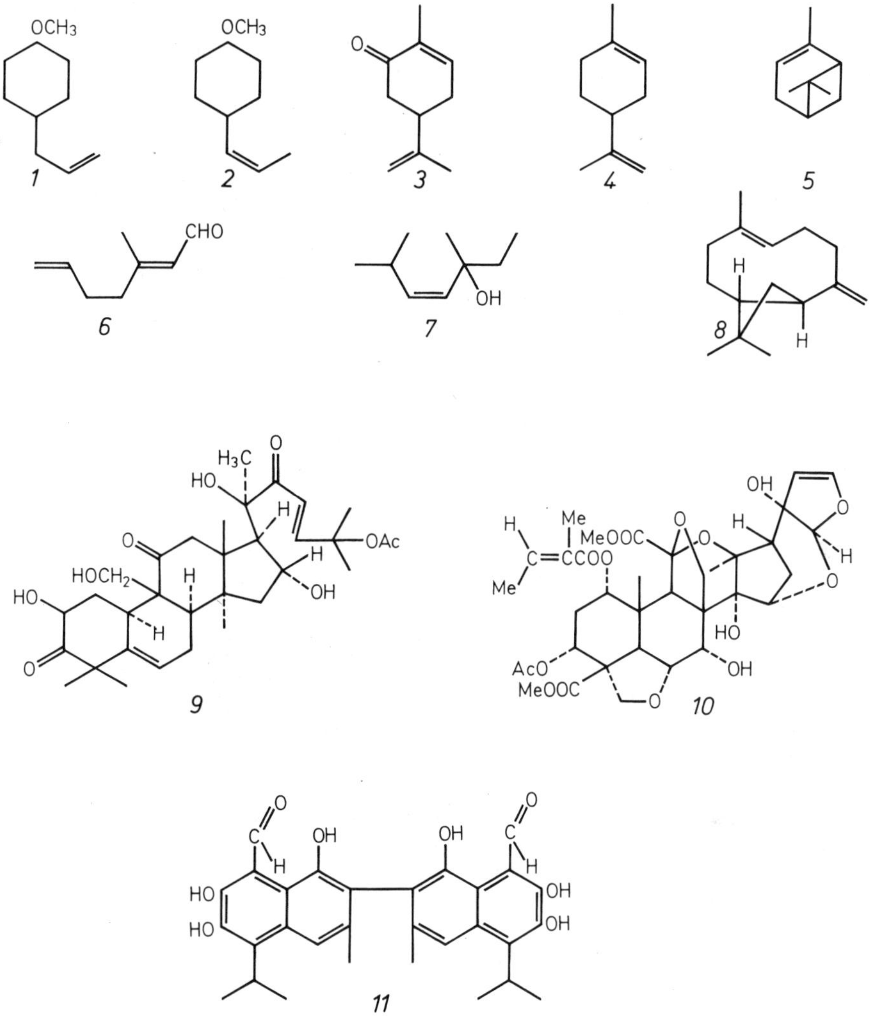

Fig. 9. Structural formulae of some isoprenoids. *1* Methyl chavicol, *2* Anethole, *3* Carvone, *4* Limonene, *5* α-Pinene, *6* Citral, *7* Coriandrol, *8* Caryophelene, *9* Cucurbitacin A, *10* Azadirachtin, *11* Gossypol. Taken from [12]

more C_5 isoprenoid units, and range from C_{10} compounds, monoterpenes, to poly-isoprenoids, such as rubber [16]. The monoterpenes, which are more or less absent in lower land plants, e.g., ferns, horsetails and lycopods, are especially important as antifeedants in higher plants. Sesqui- and di-terpenes show their first striking diversity among angiosperms, and are widely involved in such plants as antifeedants (e.g., gossypol and azadirachtin, Fig. 9).

The alkaloids, highly toxic nitrogenous bases, are further evidence of the evolution in higher plants of secondary compounds which have antifeedant activities to many phytophagous insects (Table 2). Alkaloids (Fig. 10) apparently are a special angiospermous trait, and they arise from many different precursors [16]. Lower plants

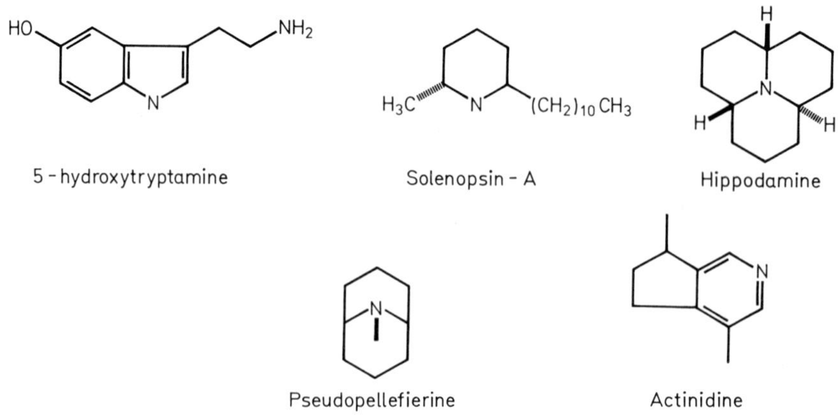

Fig. 10. Representative nitrogenous bases found both in animals and plants which are antifeedants to some insect species

certainly have nitrogenous antifeedant compounds, e.g., cyanogenetic glycosides in ferns, but apparently lack true (i.e., aromatic) alkaloids. Though alkaloids are allomonic to many insects, their many-fold greater toxicity to numerous mammalian herbivores has been interpreted by some investigators as evidence that angiosperms primarily evolved these nitrogenous compounds as defenses against mammals.

Flavonoids (see quercetin for some basic structural characteristics) are another major class of chemicals in plants which may be antifeedants to given insects.

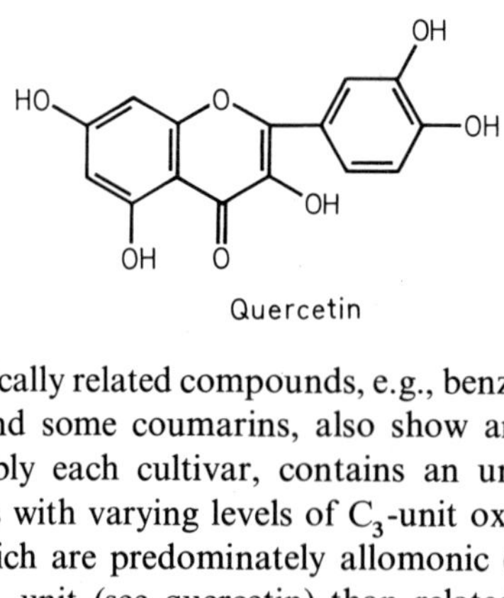

(8)

Quercetin

Biosynthetically related compounds, e.g., benzophenones, lignans, condensed tannins, stilbenes and some coumarins, also show antifeedant actions. Each plant species, and probably each cultivar, contains an unique array of flavonoids and related compounds with varying levels of C_3-unit oxidation (unsaturation) (Table 3). Compounds which are predominately allomonic (e.g., antifeedant) usually have a more oxidized C_3 unit (see quercetin) than related compounds (see catechin) which are more usually kairomonic.

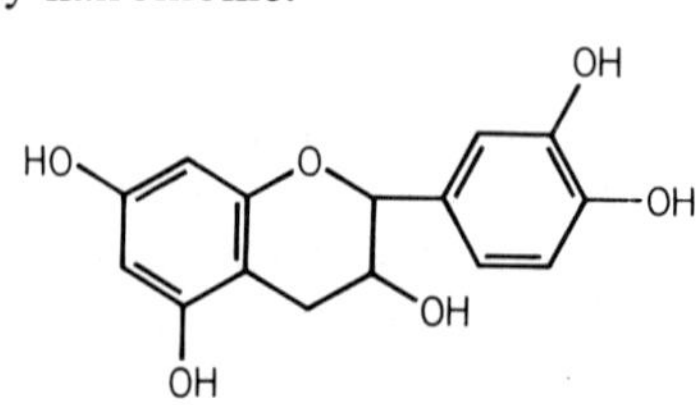

(9)

Catechin

Table 3. Oxidation states in the C_3 unit of different types of flavonoids

Flavonoid name	Structure of C_3 unit
Flavan-3-ols	A—CH_2—CHOH—CHOH—B
Hydrochalcones	A—CO—CH_2—CH_2—B
Chalcones	A—CO—CH=CH—B
Flavanones	A—CO—CH_2—CO—B
Leucoanthocyanidins	A—CHOH—CHOH—CHOH —B
Flavones	A—CO—CH_2—CO—B
Anthocyanidins	A—CH_2—CO—CO—B
Benzalcoumaranones	A—CO—CO—CH_2—B
Flavanonols	A—CO—CHOH—CHOH —B
Flavonols	A—CO—CO—CHOH—B

[a] Taken from Geissman [25)]

This situation means that based solely on flavonoid chemistry each species and cultivar of higher plant may be expected to demonstrate unique antifeedant activities.

If all classes of secondary compounds in a plant species are considered, then each species surely provides an unique spectrum, or concentration, of antifeedants.

5.2 Specific Chemicals

Perhaps the most widely known phytochemical with antifeedant activities to numerous insects is the triterpenoid, azadirachtin [14, 27)]. It is present in the leaves and berries of *Azadirachta indica* (Indian neem tree) and *Melia azedarach*. The trees are found commonly in India, and East and West Africa.

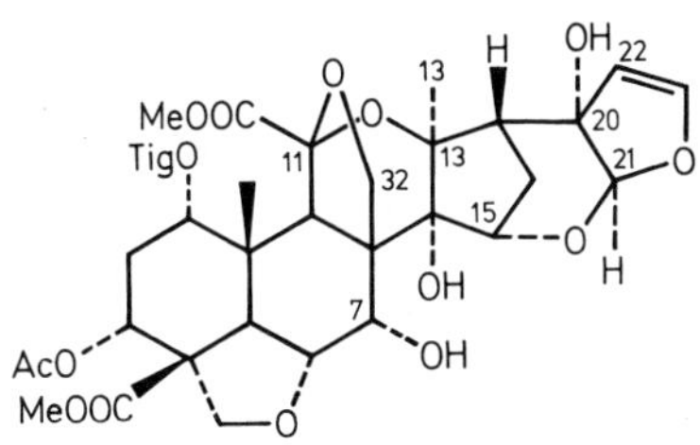

(10)

Azadirachtin at 40 μg/liter or 1 ng/cm² of filter paper has caused 100% inhibition of desert locust (*Schistocerca gregaria*) feeding. It however lacks significant antifeedant activities against North American grasshopper species. Other insect pests, e.g. *Oryzaephilus surinamensis* (a pest of stored grains and seeds), also cope readily with azadirachtin and may damage 100% of neem kernels. Thus, as with all compounds, azadirachtin is not an antifeedant to all insect species.

Although the complex structure ($C_{35}H_{44}O_{16}$) of azadirachtin prevents its synthesis in commercial quantities, its abundance (e.g., 800 mg/300 g) in neem seeds and the ease of neem tree cultivation both encourage the growth of such trees as commercial sources of this chemical.

Though widely publicized for its biological activities, azadirachtin is only about one-half as potent as the terpenoid derivative, warburganal (see Sect. 4), to *Spodoptera exempta* [14]. Warburganal occurs in the bark of *Warburgia ugandensis*, and is found with the closely related compounds, polygodial and ugandensidial, in other *Warburgia* species. All three compounds show antifeedant properties, but warburganal is, by far, the most active (e.g., 0.1 ppm against *S. exempta*) [14]. Another terpenoid derivative, the

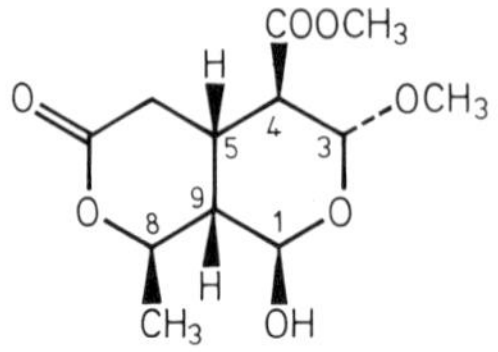

Xylomolin

(11)

secoiridoid, xylomolin, from the unripened fruits of *Xylocarpus moluccensis* also shows moderate antifeedant action to *S. exempta* at 100 ppm [28].

Several sesquiterpene hydrocarbons, caryophyllene, α-selinene, β-selinene and β-copaene, in the legume, *Hymenaea courbaril*, are antifeedants to the generalist feeder, beet armyworm (*Spodoptera exigua*) [32,33]. The sesquiterpene lactone, parthenin, in the plant *Parthenium hysterophorus* is an antifeedant to a diverse group of insects including *Dysdercus koenigi*, *Aedes aegypti*, *Tribolium castaneum*, *P. americana* and *Phthormea operculella* [31,32]. The germacrane sesquiterpenes, shiromodiol monoacetate and shiromodiol diacetate, in the leaves of *Parabenzoin trilobrum* inhibited tobacco cutworm (*Spodoptera litura*) feeding [29].

Some species of the genus *Vernonia* produce a germacranolide-type sesquiterpene lactone, glaucolide-A, with known antifeedant actions against some species of Lepidoptera [30].

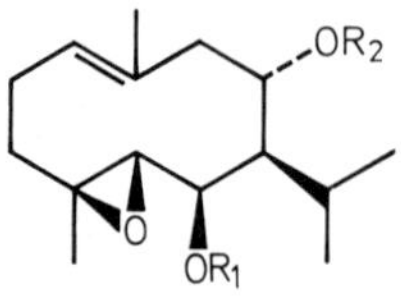

(12)

Shiromodiol diacetate. $R_1 = R_2 - Ac$
Shiromodiol monoacetate. $R_1 = H; R_2 = Ac$

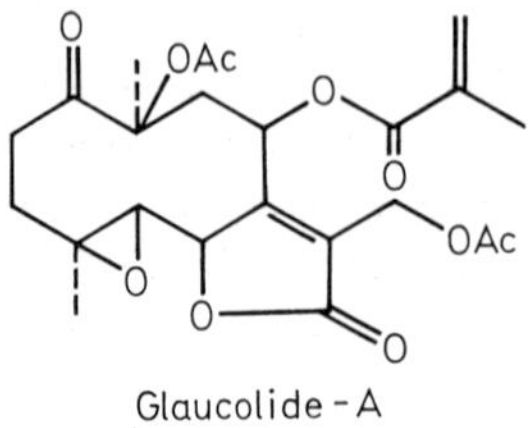

Glaucolide-A

(13)

However, cabbage looper (*Trichoplusia ni*) and the yellow woollybear (*Diacrisia virginica*) are sympatric with species of *Vernonia*, and when they were given a choice of species with, versus without, glaucolide-A they tended to prefer the former. The saddleback caterpillar (*Sibine stimulea*) that naturally feeds on the glaucolide-A lacking *Vernonia flaccidifolia* prefers the latter. Glaucolide-A is also an antifeedant to two generalist lepidopteran feeders, the fall and southern armyworms (*Spodoptera eridania* and *S. frugiperda*). A concentration of approximately 1% glaucolide-A gave maximal inhibition in bioassays, and this is about the quantity of chemical in many sesquiterpene lactone-containing Compositae. These overall findings with glaucolide-A and various Lepidoptera clearly emphasize the strikingly different coevolutionary situations which commonly develop between various species in a genus of plants and several species in a grouping of insects [30]. A similar dichotomy of situations exists with *Gossypium* plants (e.g., cotton) and several insects. Some sesquiterpenoids (e.g., gossypol and related compounds) in such plants excite feeding by the boll weevil (*Anthonomus grandis*), but they apparently are repellents and antifeedants to the lepidopterans, *Heliothis zea* and *Heliothis virescens*, and blister beetles in the genus *Epicauta* [34].

Several clerodane diterpenes from plants in the Verbenaceae are proven antifeedants to several insects [29]. The chemicals include clerodendrin-A, clerodendrin-B, clerodin, caryoptin, dihydrocaryoptin, caryoptin hemiacetal, caryoptinol, 3-epicaryoptin, 3-epidihydrocaryoptin, dihydroclerodin-1, clerodin hemiacetal and phytol. Except for phytol, all at 50–200 ppm competely inhibited *S. litura* larval feeding within 2 hour. Clerodendrin-A and -B also prevented the feeding of oriental tussock moth larvae, *Euproctis subflava*, at 1000 ppm and European corn borer larvae, *Ostrinia nubilalis*, at 5000 ppm, but not the feeding *Calospilos miranda* larvae at less than 5000 ppm. These results suggest that such diterpenoids are more effective as antifeedants against polyphagous rather than monophagous insects. The grayanoid diterpenes, grayanotoxin III and kalmitoxin I and II, likewise inhibit the feeding of the very polyphagous gypsy moth (*Lymantria dispar*) larva. Leaves of *Isodon inflexus* contain inflexin and those of *I. shikokianus* var. *intermedius* yield isodomedin. These *end*-kaurenoid diterpenes are both antifeedants and cytotoxins to *S. exempta*.

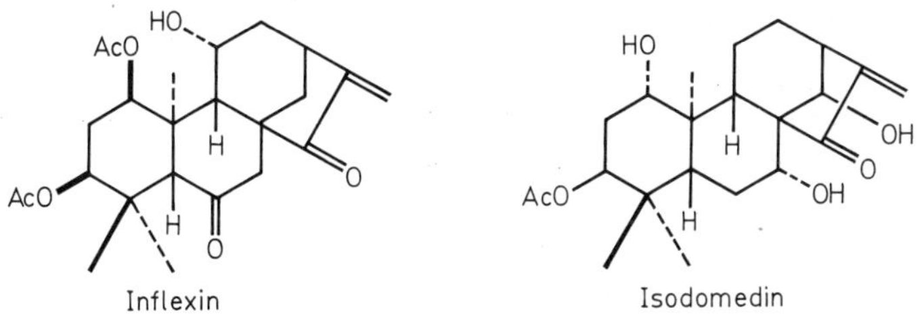

(14)

In addition to the previously discussed azadirachtin, numerous other plant-derived triterpenoids are antifeedants to some insects. The leaves of the shrub *Harrisonia abyssinica* contain the pentacyclic harrisonin which has antifeedant properties to insects, as well as many uses in East African folk medicine. It is an effective antifeedant against *S. exempta* at 20 ppm [14].

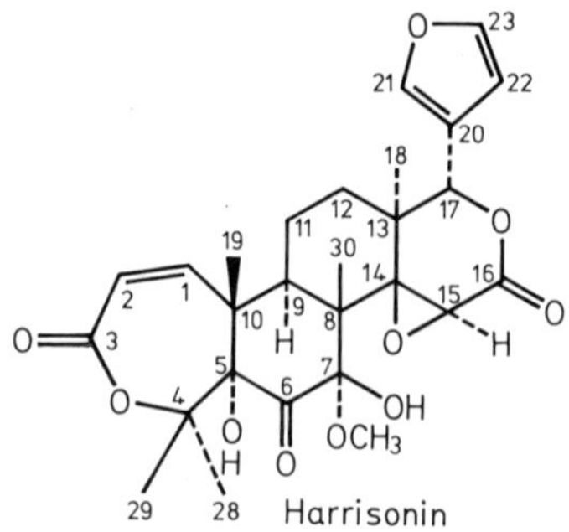

(15)

Meliantriol, a hydroxylated steroid, has proven antifeedant activity to *Schistocerca gregaria*. Friedelin inhibits feeding by some insect species. Glycosides of some steroidal alkaloids, e.g. demissine, solacauline, tomatine, leptine I and leptine II, found in Solanaceae variously reduce feeding by the Colorado potato beetle, *Leptinotarsa decemlineata* [18, 19, 24].

Several groups öf acetogenins are historically important phytochemicals with demonstrated antifeedant properties against some insects. Resistance of some maize varieties to European corn borer larvae involves the antifeedant activities of acetogenous benzoxazolinone compounds (e.g., DIMBOA) in especially young leaves [4]. The acetogenin, 5-hydroxy-1, 4-naphthoquinone (juglone), in *Carya* (hickory) trees is a potent antifeedant to *Scolytus multistriatus* (the smaller European elm bark beetle) that naturally feeds on *Ulmus* trees, but does not have such allomonic effects on *S. quadrispinosus* (the hickory bark beetle) that naturally feeds on *Carya* [7, 8]. Thus, two closely related *Scolytus* species exhibit opposite coevolved reactions to this very biologically active phytochemical. This compound, juglone (see Sect. 3), is such a relatively effective, evolved defensive (e.g., antifeedant) chemical that man chose the word as the root for the name of the family (Juglandaceae) of plants in which it especially occurs. Several other 1,4-naphthoquinones also are antifeedants to several insects [35, 36, 37]. Three anthraquinones, 2-methyl anthraquinone, 2-hydroxymethyl anthraquinone and 2-formy anthraquinone, are wood-borne antifeedants to some termites. Terpenoidal quinones in some *Gossypium* plants reduce pink bollworm (*Pectinophora gossypiella*) feeding [34]. p-Benzoquinone inhibits feeding by the smaller European elm bark beetle (*S. multistriatus*) [36].

The flavonoids, phloretin from *Malus pumila*; 2′,6′-dihydroxy-4′-methoxy dihydrochalcone from *Populus deltoides*; kaempferol from *Robinia pseudoacacia* and quercetin (see Sect. 5.1) from *Quercus macrocarpa*, are antifeedants to *S. multistriatus* adults [26]. Other products of phenylpropanoid metabolism in *Parabenzoin praecox*, (+)-epieudesmin and (+)-eudesmin; and in *Piper futokazura*, isoasaron and piperenone, are antifeedants to *S. litura* larvae [29]. Coumarin in Cruciferae and Umbelliferae inhibits ingestion by curculionid beetles [12].

Other alkaloidal compounds with proven antifeedant actions include capaisin and nicandrenone in Solanaceae against chrysomelid beetles [12], and isoboldine in *Cocculus trilobus* for *Abraxas miranda* [29].

The mustard oil glycoside, sinigrin, in Cruciferae and Umbelliferae inhibits aphid feeding [12]. 2-Phenylethyl isothiocyanate in turnip (*Brassica rapa*) and 5-allyl-1-

methoxy-2,3-methylene-dioxy benzene from parsnip (*Pastinaca sativa*) inhibit *Drosophila melanogaster* feeding [38].

5.3 Induced Inhibitors

Some plants under attack by insects can decrease the aceptability of their tissues as food. They do this especially by either increasing their content of allomones (e.g., antifeedants) and/or decreasing their content of kairomones (e.g., feeding excitants). Pest attacks on some plants quickly result in changed chemical profiles in, or near, the wounded tissues [39]. This action may especially arrest a current attack. Insect herbivores also may result in attacked plants releasing systemic "signal" chemicals from the injury site for transport to unwounded tissues that are remote (e.g., at least several centimeters) from the primary infliction [39]. Such systemic effects thus may increase defenses against persistent or future insects.

Evidence strongly indicates that fragments of plant cell walls, released at the wounding site, are informational "signal" messengers that activate nearby and remote cells to strengthen biochemically their defenses against feeding [40,41,42]. The signals may be macromolecules or much smaller oligosaccharide fragments.

Known induced antifeedants in plants include proteinase inhibitors, phenolics, HCN, phytoalexins, coumestrol, cucurbitacins, ipomeamarone and quinone (Table 4) [46].

Table 4. Representative allomonic (e.g., antifeedant) reactions induced in plants by herbivory

Induced chemical product	Plant	Ref.
Cucurbitacins	Squash	43)
Coumestrol	Alfalfa	44)
Ipomeamarone	Sweet potato	45)
Phenolics	Cotton	46)
Phenolics	Chinese cabbage	47)
Quinones	Sugar beet	48)

5.3.1 Proteinase Inhibitors

Proteinase inhibitors in plants commonly are polypeptides or certain phenolics. The majority of proteinase inhibitors in plants are specific for the serine class of proteinases (e.g., trypsin and chymotrypsin) which are major food protein — digesting enzymes in insects [39]. Such allelochemics thus can indirectly reduce feeding (ingestion) through directly reducing digestion. It is well established that inhibition of insect feeding may occur when the digestive tract fills and stretch receptors in the wall of the alimentary canal fire neurons which function in feedback neuronal loops which inhibit further ingestion.

These inhibitors are widely distributed in the plant kingdom, and especially in seeds and tubers [49]. They were first discovered in tubers of potato (*Solanum tuberasum*), and the wound-induced accumulation of such inhibitors was studied first in tomato

(*Lycopersicon esculentum*) leaves [39]. Two inhibitors, I and II, are found in both potato and tomato leaves. Inhibitor I has a molecular weight of 41,000, and is a pentamer of 8,000 dalton subunits. The molecular weight of inhibitor II is about 23,000.

The presence of inhibitors in potato tubers is apparently regulated by physiological signals that are under developmental control; whereas, the quantity of inhibitors in either potato and tomato leaves is regulated by chemical signals (i.e., proteinase-inhibitor inducing factor, PIIF) released by stimuli from the environment [39]. The quantity of PIIF released at, or near, the wound site depends upon the location and severity of the injury. This factor is transported, apparently in the phloem, to unwounded leaves within 2–3 hours. Transport is directed primarily upward to younger leaves [39]. PIIF activity has been consistently associated with oligo- and polysaccharide fractions of 5,000 to 10,000 molecular weight. It is known to be a highly methylated polysaccharide containing galacturonic acid, rhamnose, galactose, arabinose and fucose. Subsequent research has shown that various polysaccharide fractions from plant cell walls are as active as PIIF from tomato [39]. However, not all pectic polysaccharides show PIIF activity.

It is now hypothesized that pectic oligosaccharide fragments with PIIF activity are released (i.e., decompartmentalized) at the site of injury (e.g., insect feeding wound) through degradation of the plant cell wall by hydrolytic enzymes unleased either from ruptured plant cells or from the invading pest (e.g., insect). The transported PIIF apparently reacts with specific receptors in unwounded cells of the plant to induce the synthesis and accumulation of proteinase-inhibitor proteins which may reduce insect feeding.

5.3.2 Phytoalexins

Classical phytoalexins are defensive compounds induced in plant tissues near sites of attack by pathogenic organisms. They represent a diversity of low molecular weight products in chemical classes including sesquiterpenes, diterpenes, polyacetylenes, dihydrophenanthrenes, stilbenes and isoflavonoids. Their production has been especially investigated in the Leguminosae and Solanaceae, but has been demonstrated in more than 75 species in at least 20 botanical families [50]. With the exception of *Ginkgo*, phytoalexins have been discovered mostly in angiosperms and in a few conifers.

Phytoalexins are produced in plants in response to the release of elicitors (e.g., PIIF) at sites of primary plant injury (e.g., fungal hyphal infection). Some phytoalexins are allomonic to some insect herbivores. Thus, such defensive compounds induced in a higher plant through invasion by one agent (e.g., a fungus) may also readily reduce the injury level subsequently inflicted by a second agent (e.g., herbivorous insect).

Phytoalexins apparently are only one major component in multi-factor plant responses to pathogens and pests. Other components include lignification, suberization, callose formation and productions of agglutinins and inhibitors of extracellular hydrolases.

5.3.3 Phenols and Quinones

Aromatics which undergo reversible oxidation-reduction constitute another group of inducible phytochemicals which may exhibit striking antifeedant activity. A major component of the induced wounding reaction in plant tissues involves the various degrees of oxidation of phenols which are decompartmentalized (i.e., exposed to atmospheric oxygen and/or plant or herbivore phenoloxidases) by injury to plant cells [51]. Many of the resultant quinoid oxidation products are antifeedants to numerous insects. A classic example is the oxidation of p-hydroquinone (a feeding excitant for *S. multistriatus*) to its redox couple, p-benzoquinone, which is an antifeedant for this elm bark beetle [35,36], as follows.

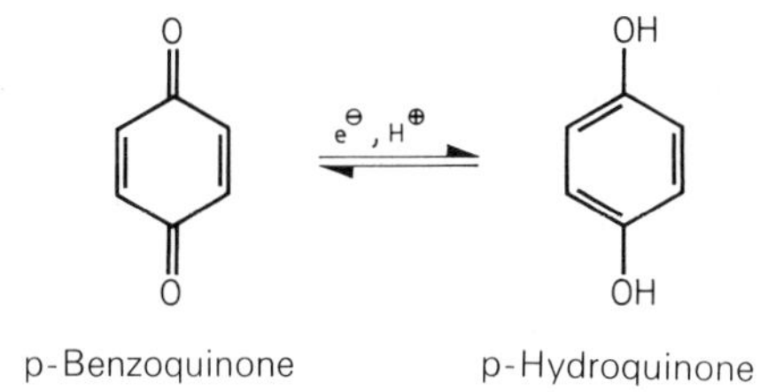

(16)

6 Antifeedants in Animals and Humans

As with phytophagous insects, those that feed on animals and humans depend on an array of physical and chemical stimuli in identifying a host [53]. Feeding deterrents and inhibitors (i.e., antifeedants) apparently are more important than attractants and feeding excitants in insect decisions regarding animal host versus non-host situations [53]. Thus, allomones, including antifeedants, apparently are similarly important in insect-animal host interactions and insect-plant host relations [54]. Chemicals associated with the animal's skin are especially important as antifeedants [53]. With hematophagous species, the chemistry of the blood also may influence feeding activity. During initial feeding trials, a hematophagous insect may inject antigens which induce an immunological reaction in the host's blood. Such reactions may change the animal's status to non-host with regard to the initially feeding insect species.

Chemicals which, among other activities, are known to inhibit, or prevent, insect feeding on animal hosts constitute a diverse group of compounds [54,55]. Most of this knowledge has been obtained from investigations of insects as hosts and non-hosts. Such information involving other animals and humans is very sparse [53,54]

Antifeedant compounds include esters, aldehydes, alcohols, ketones, quinones, phenols, lactones, sulfides, nitros, furans, alkaloids, ethers and various hydrocarbons. Among the esters, compounds range from C_5–C_{16}, but most are C_8–C_{18}. Nearly 50% are acetates; and even numbered, aliphatic acids predominate. Among the aldehydes most are aliphatics and nearly 50% have α, β-conjugated structure. Terpenoidal aldehydes are quite common. Nearly 66% of the alcohols with antifeedant effects are C_{10}–C_{20} compounds. More than one-half are primary alcohols, and about 25% are terpenoids. Ketones range from C_4–C_{22}. Methyl and ethyl ketones are common, but aromatic structure is rare. Quinones include 1,4-benzoquinones and 1,4-naphthoquinones. Some phenols are reported as antifeedants, but that allomonic activity frequently may be more directly attributable to their redox couples, quinones. Lactones are common antifeedants produced by insects. Among the hydrocarbon antifeedants, over 25% are methyl-branched terpenes. Most are saturated, but a large variety of alkenes occur. Terminal unsaturation has been reported in about one-fourth of such hydrocarbons [55].

Many of the antifeedants reported from animals also occur in plants [37]. Thus, antifeedant activities in plants and animals commonly involve very similar, or even the same, chemical structure-biological action relationships. Benzoquinones, e.g. 1,4-benzoquinone, are a major example of such antifeedant chemistry associated with both animals and plants.

7 Structure-Activity Relationships

Because the antifeedant actions of molecules, ions or free radicals are only effective against some, but not all, insect species, it remains unwise to make absolute statements regarding relationships between such antifeedant activity and chemical structure. However, if one examines the structures of a large number of molecules which are especially known as insect antifeedants in comparison to those of a group of compounds which are predominantly considered as feedants, one is likely to find that those with antifeedant activity possess the more oxidized (or unsaturated) structures. Many oxidizing agents thus are antifeedants to one or more insect species. An example of such a relationship between the relative oxidation level in a compound and its antifeedant action involves p-benzoquinone of the redox couple, p-hydroquinone: p-benzoquinone, as follows, and *S. multistriatus* [35,36].

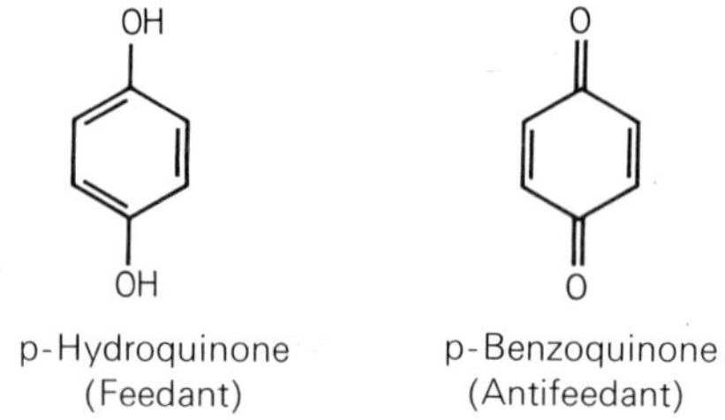

(17)

Our best understanding of the characteristics which make p-benzoquinone an antifeedant to this beetle indicates that its α,β-conjugated (unsaturated) structure is very important. These characteristics enable p-benzoquinone, but not p-hydroquinone, to react as an electrophilic substance with a nucleophilic terminus (e.g., sulfhydryl) in a receptor macromolecule or binding site, as follows

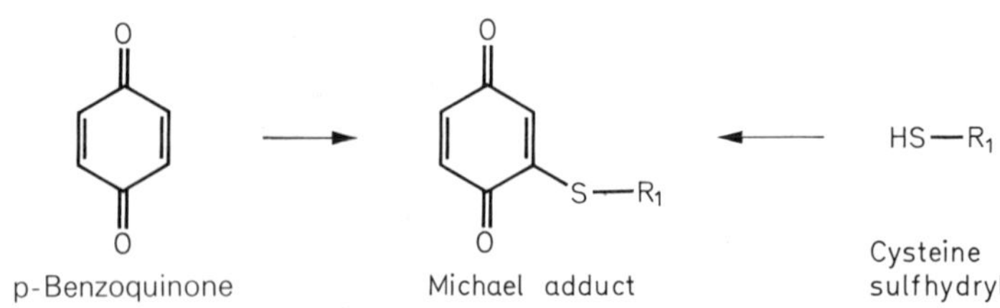

(18)

As with p-benzoquinone, the high antifeedant activity of warburganal against *S. exempta* apparently is caused by its electrophilic ability to form a Michael addition with a nucleophilic (e.g., sulfhydryl) terminus in the receptor [14]. Other structure-activity studies[14] with the related polygodial indicated that its antifeedant action against *S. exempta* was lost when one or both of its aldehydes were reduced to hydroxyls or oxidized to carbonyls. Thus, the oxidation state of key functional groups and of the total molecule is critical to antifeedant action.

The potent antifeedant properties of 5-hydroxy-1,4-naphthoquinone, juglone, and other 1,4-naphthoquinones likewise are attributed to their electrophilic and oxidative characteristics [35, 36, 56–59]. They are known to form a Michael addition with nucleophilic sulfhydryls of the redox-complex (receptor) macromolecule in the dendritic membrane of *P. americana* chemosensory neurons, as follows

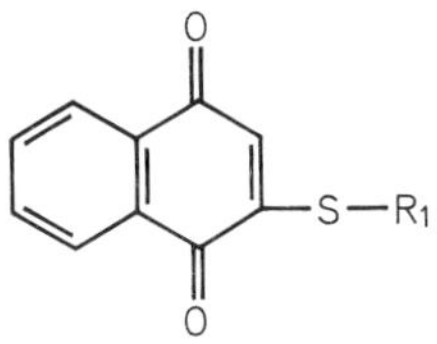

(19)

Michael adduct

8 Perception of Antifeedants

Perception of a molecule, ion or free radical as an antifeedant is ultimately expressed behaviorally. The observed behavior is an inhibition, or reduction, of feeding. Our understanding of the involved chemical senses, whether of insects, other animals or humans, has been quite limited. The resultant (efferent) behavioral parameters of antifeedant perception are relatively well understood, and reasonable experimental methods for quantifying such behavior are plentiful. The adult blow fly (*Phormia regina*) assay involving the withdrawal of an extended proboscis upon the fly making direct contact with an antifeedant, via either its proboscis or tarsus, is a classic example of such an experimental model [60, 61]. Various assays which provide a feeding choice between at least two treatments, e.g., feeding excitant versus feeding excitant + candidate antifeedant, presented on a standardized "neutral" substrate (e.g., elderberry pith disc), also readily visualize and allow quantification of the behavioral components of antifeedant perception [7, 14, 26]. However, the afferent (e.g., sensory) events involved in the cause-effect neural interactions of antifeedant perception still have been considered by many experts as part of a "black box" [62, 63]. But significant progress toward defining and understanding the unifying principles of the energy transduction which occurs between an antifeedant and dendrites of appropriate chemosensory neurons has occurred especially since the late 1960's. Much of the progress has resulted from investigations of American cockroach (*P. americana*) perception of the antifeedant properties of 1,4-naphthoquinones [3, 56, 57, 58, 64, 65, 66].

8.1 Perception of Naphthoquinones as Antifeedants

Behavioral analyses of the antifeedant effects of several 1,4-naphthoquinones on adult *S. multistriatus* [35, 67] and adult *P. americana* [58] indicated that the order of relative feeding inhibition was 5-OH > 1,4- > 2-OH > 2-CH_3. This order of behavioral effects could be largely attributed to the relative combined redox potential and hydrogen bonding capabilities of these compounds [35, 36]. Because these naphthoquinones are so readily capable of redox reactions [35, 36], dropping-mercury-electrode polarography was utilized extensively in quantitating electrochemically (e.g., using E 1/2 values) their energy exchange (i.e., energy transduction) with the redox-complex lipoprotein (i.e., receptor) which was extracted from the plasma membrane of chemosensory neurons in sensilla on the cockroach's antennae [58, 59, 68)]. The redox-complex lipoprotein [56, 66, 68)] which binds, and exchanges, energy with the antifeedant naphthoquinones was found to contain cysteine residues bearing sulfhydryl groups which form Michael-type (i.e., 1,4-addition) reactions with, and reduce, the antifeedants [56, 57, 58, 65, 66)]. Thus, the redox-complex lipoprotein is a sulfhydryl-containing macromolecule which reacts readily with naphthoquinones by a combination of Michael addition and redox interactions. The primary energy transduction which occurs between these antifeedants and the redox-complex macromolecule in the chemosensory neuron thus may be removed from classification as a "black box". Much subsequent research on pharmacologically altering organismal perceptions of allomones (e.g., antifeedants), kairomones or pheromones based on sulfhydryl-disulfide protein chemistry has strongly supported the above interpretation of the energy transduction which is involved in antifeedant perceptions [14, 57, 68, 69, 70, 71)].

8.1.1 Ultrastructural and Cytochemical Parameters

The ultrastructural and cytochemical constraints of the *in vivo* chemoreception system must be accommodated by any physicochemical interpretation of the energy transduction involved in perceptions of antifeedants [70)]. Antifeedants make contact with the organism's sensory neurons through the ultrastructure and associated cytochemistry of sensilla on especially the mouthparts, antennae or tarsi. In insects, sensilla trichodea, frequently having a single terminal opening (uniporous) to connect the external

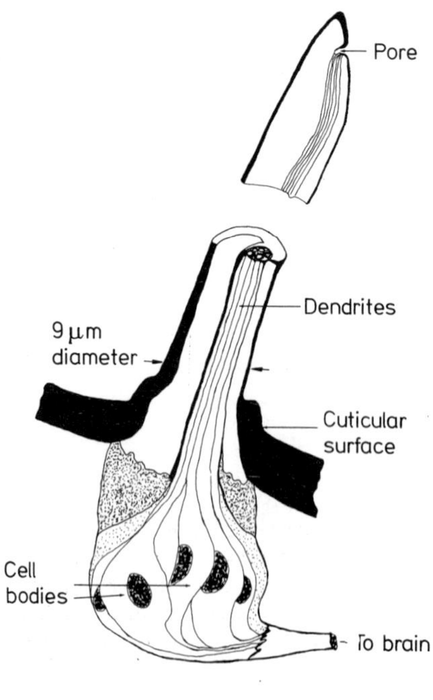

Fig. 11. Schematic diagram of the internal structure of a single blowfly (*P. regina*) uniporus taste (contact) sensillum (adapted from [60])

environment with the sensory neurons in the lumen of the sensillum (Fig. 11), function especially as contact chemoreceptors. They thus are important sensory structures in decisions regarding gustation and ingestion. Stimulants may readily enter the lumen of this type of sensillum and directly exchange energy with redox-complex, energy-transducer macromolecules in the dentritic membranes of the exposed sensory neurons [67,72].

Though multiporous sensilla are generally considered as especially functional in the detection of volatile (olfactory) stimulants, such sensilla basiconica, with their thousands of cuticular perforations, also function through contact as "taste" receptors. In the case of sensilla basiconica (Figs. 8, 12), stimulants may gain direct access to associated chemosensory neurons via the cuticular pore-pore tubule route to the dendrite [67,70,72]. Though stimulants, whether excitants or inhibitors, may penetrate the cuticle of either sensilla trichodea or sensilla basiconica and thus gain direct access to the associated sensory neurons, it is not established whether such direct contact between stimulant and nerve is necessary for energy exchange [56,57,66]. Antifeedants probably can effectively exchange (or share) energy indirectly with the chemosensory neurons through just binding (complexing) with proteinaceous contents (Fig. 13) shown to occur on the external surface of chemosensory sensilla; in the lumen of cuticular pores and pore tubules of sensilla; and in the "sensillum liquor" [66,73,74]. These saline-soluble sulfhydryl proteins bind naphthoquinone feeding inhibitors [56,65,66,74] as well as excitants including pheromones. Among other functions, they

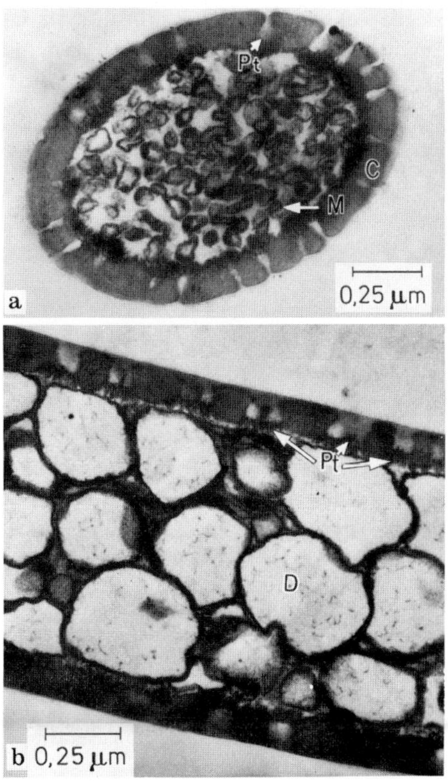

Fig. 12a and b. Ultrastructure of a multiporus chemosensitive sensillum basiconicum on the antenna of *S. multistriatus*. **a** Cross section showing flask-shaped pores through the cuticle (C) of the sensillum, plasma membranes (M) of nerve dendrites and pore tubules (Pt) which connect each cuticular pore to a dendritic membrane. **b** Longitudinal section of such a sensillum. Dendrite (D), pore tubule (Pt)

thus may serve to retain temporarily on the external cuticular surface messenger molecules, ions and free radicals which play roles in the organism's decisions about antifeedants.

Electrophoresis of the contents of 0.9% NaCl, potassium phosphate buffer, pH 7.0, after dipping intact antennae of live *P. americana* in the solution, yielded two sharply resolved proteinaceous bands [66, 73, 74]. These proteins incorporated ^{35}S-cystine *in vivo* with a metabolic life between 9–23 days [74]. Such ^{35}S incorporation was considered important to understanding the functioning of these proteins in chemoreception processes because the studied 1,4-naphthoquinone antifeedants were already known to react selectively with a sulfhydryl-rich lipoprotein energy-transducing receptor in the dendritic membrane of the involved chemosensory neurons [56, 59, 65, 66, 75,]. Thus it is reasonable for sulfhydryl-rich saline-soluble proteins, in the contents of cuticular pores and secreted on the exposed surface of chemosensory sensilla, to interact effectively with especially the 1,4-naphthoquinone antifeedants. Cleaving the sulfur atoms of sulfhydryls in such proteins by Raney nickel treatment [76] significantly reduced their binding of the antifeedants [65]. This treatment also destroyed the energy-transducing capabilities of the purified sulfur-rich lipoprotein from the dendritic membrane of *P. americana* chemosensory neurons [3, 56, 65, 66].

The two electrophoretically resolved bands of saline-soluble, sulfhydryl-rich proteins which are in the pore tubule-cuticular pore contents and on the surface of the cuticle are also prominent components of haemolymph proteins of *P. americana* [73, 74]. Proteins in the haemolymph of insects may become transferred by endocytosis into the accessory (i.e., trichogen, tormogen and sleeve) cells associated with the ultrastructure, cytochemistry and overall function of chemosensory sensilla [70, 77]. Such proteins

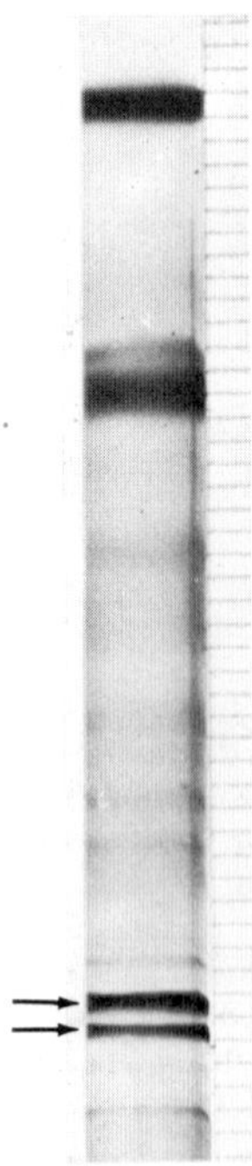

Fig. 13. Disc gel electrophoretic resolution of saline-soluble "interfacin" proteins which bind antifeedant 1,4-naphthoquinones; here in haemolymph proteins from male adult *P. americana* [see [66,73]]

subsequently may be secreted from the accessory cells into the ultrastructural cavities in chemosensory hairs [70]. Once in the "sensillum liquor" of such chemosensory sensilla, such proteins may be moved readily to the contents of the directly contacted pore tubules, and then be secreted via positive sensillum liquor-based pressure onto the cuticular surface of chemosensory sensilla (i.e., sensilla trichodea and sensilla basiconica [70]. This combined ultrastructure, cytochemistry and secretion thus may interface electrochemically the insect's chemosensory neurons with the external environment via a dynamic system of sulfhydryl-disulfide proteins [57]. It is well established that such proteins may function as redox complexes and biological amplifiers; and as such, respond reversibly and sensitively to changing oxidation states in the environment [57, 78].

Living cells must be protected from oxidizing conditions (e.g., an excess of electrophilic substances such as 1,4-naphthoquinone antifeedants) if life is to be sustained [57]. Every cell and organism thus must continuously monitor oxidation states in its surrounding environment. The above described "interfacing" system of sulfhydryl-disulfide polypeptides in the chemosensory mechanisms of *P. americana* constitutes such a continuous monitor of environmental oxidation states [57]. As discussed earlier in this chapter, antifeedant properties are especially associated with relatively oxidized compounds; thus, a sensory system based especially on the detection of changing oxidative states provides an effective mechanism for perceiving antifeedants and many other substances possessing allomonic activities. The fact that such a redox-based chemosensory, energy-transduction system between the external environment and the sensory neural membrane is readily reversibly mechanistically (i.e., oxidative versus reductive) enables it to exchange energy also with (detect) compounds (e.g., relatively reduced molecules) with especially kairomonic behavioral effects. This interpretation is compatible with most facts regarding energy-transduction aspects of chemoreception by insects and other organisms; and with previously proposed oxidative-reductive (yin-yang) mechanisms for such chemosensory perceptions [3, 35, 57, 58, 59].

The energy units for such a redox-based energy-transduction system are hydrogens [79]. Thus, varying gradients of hydrogens and electrons may be created across the continuum of sulfhydryl-disulfide (redox) proteins extending from those exposed on the cuticular surface, through those in the cuticular pore-pore tube channel, in those of the sensillum liquor and into the redox-based macromolecular transducer which is an integral component of the dendritic membrane of the chemosensory nerve cell [3, 57]. These dynamic physicochemical properties of the sensillum-associated chemosensory interface between the insect's nervous system and the external environment enable net energy changes, attributable to such hydrogen and electron gradients, to be translated (transduced) directly into a protonic potential in the dendritic membrane [80, 81]. The resultant potential makes the membrane either more or less polarized. This overall change in potential either increases or decreases the rate at which a threshold generator potential may be developed in the dendritic membrane and at which the chemosensory neuron thus may be fired [57, 82].

Because the above described sulfhydryl-disulfide polypeptides make a critical functional interface between the insect's chemosensory neurons in the sensillum and the external environment, these proteins are now being named functionally as *interfacins*.

8.1.2 Energy Transduction in the Dendritic Membrane

Energy transduction aspects of antifeedant chemoreception have been especially investigated with 1,4-naphthoquinones interacting with *S. multistriatus* and *P. americana* [3, 7, 35, 56, 57, 58, 59, 65, 66, 68]. The understanding which has emerged is highly compatible with Nobel laureate Peter Mitchell's chemiosmotic theory of membrane bioenergetics as applicable to such energy transduction [80, 81]. The Mitchellian theory has now been widely accepted internationally as a reasonable description of the essential components of transmembrane energy-transduction machinery. It combines two concepts of translocation; one, electron (e^-) translocation, which enables recognition and description of electromotive forces (i.e., electricity) and the other, group translocation (e.g., H), which enables the recognition and description of chemicomotive forces (e.g., protonmotive forces, proticity). Based on these concepts energy transfer (transduction) between an antifeedant and the dendrite of the chemosensory neuron may be considered as arising from thermally activated diffusion of solutes (e.g., Na^+, K^+) and chemical groups (e.g., H, OH) down the electrochemical potential gradients of their various compounds and complexes in the external (i.e., sensillum liquor, pore tubule, cuticular pore) and internal (nerve cell cytoplasm) hydrophilic (i.e., aqueous) media surrounding the dendritic membrane (Fig. 14), and in the generally hydrophobic lipid-protein intra-membrane medium itself (Fig. 14).

The physiological level of the Mitchellian chemiosmotic-coupling concept is formulated explicitly and unambiguously in terms of four fundamental postulates. These postulates correspond to the four structural and functional components illustrated in Fig. 15. The postulates are:

1. An ATP synthase is involved, and it is a chemiosmotic membrane-located reversible ATPase.
2. A chemiosmotic membrane-located redox complex having characteristic $\leftarrow H^+/2e^-$ stoichiometry is involved.

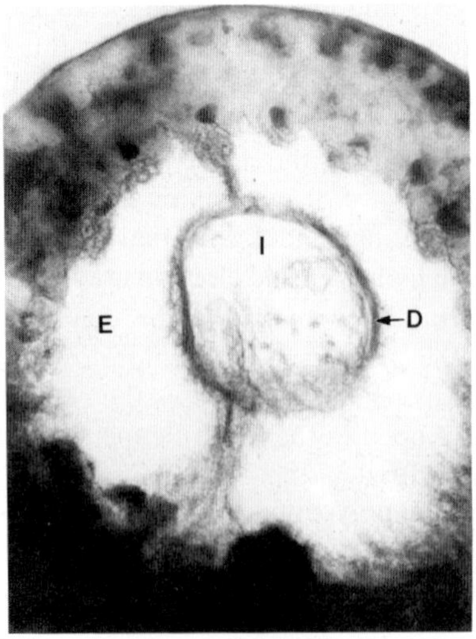

Fig. 14. High voltage electron micrograph revealing the "fuel cell" ultrastructure of a sensillum basiconicum on the antenna of American cockroach (*P. americana*). External (E) and internal (I) hydrophilic media surround the dendritic (i.e., hydrophobic) membrane (D) (60,000 ×) [see [75, 81]]

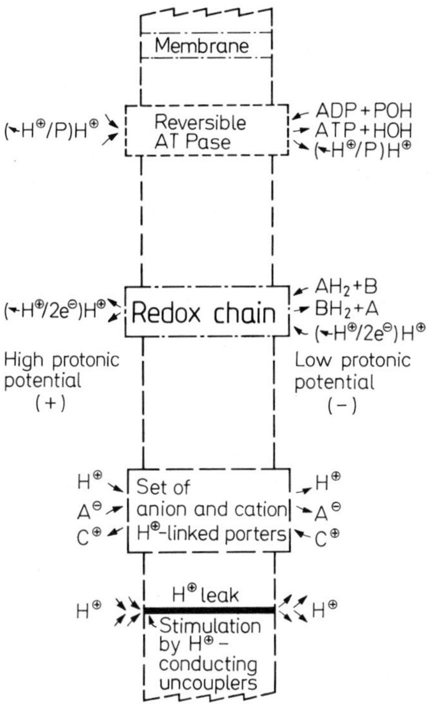

Fig. 15. Schematic of the four components of the physiological level of the Mitchellian chemiosmotic hypothesis of energy coupling in membrane bioenergetics: Reversible ATPase. Redox chain (complex). Solute porters. The hydrophobic coupling membrane [adapted from [81]]

3. Proton-linked (or hydroxyl ion-linked) solute porter systems for osmotic stabilization and metabolite transport are present.
4. Systems (components) 1 through 3 are plugged through a topologically closed insulating membrane, called the coupling membrane. It has a nonaqueous osmotic barrier phase of low permeability to solutes in general and to hydrogen ions and hydroxyl ions in particular.

The four above postulates are now widely accepted as experimentally established facts [81]. This four-component energy-transduction system thus generates proticity across the coupling membrane and this proticity energizes the aqueous conductors on either side of the membrane. The proticity-based power can be drawn off from the aqueous conductors by the other plug-through complexes in the membrane, such as the reversible protonmotive ATPase and the solute (e.g., Na^+) pumps.

Whatever living machinery such as the Mitchellian four-component energy-transduction system does, it has to pay for it in terms of energy. Hydrogens are the fuel of living systems [79]. Energy transduction thus is the orderly process by which hydrogens (energy units) are ultimately donated to, or withdrawn from, energy pools in living systems [57]. Again, major energy pools involved in the energy transduction of antifeedant chemoreception consist of (a) the combined contents of cuticular pores, pore tubules and sensillum liquor (the external aqueous medium, Figs. 12, 14, 15) and (b) the contents of the nerve cell cytoplasm (the internal aqueous medium, Figs. 12, 14, 15). In living chemoreception systems, these two major energy pools normally are separated by an insulating lipid-bilayer (i.e., dendritic) membrane. The outer pool which interfaces the organism (insect) with the external environment is characteristic ally of high protonic potential (i.e., positively charged); whereas, the inner (nerve cell cytoplasmic) pool is of relatively low protonic potential (i.e., negatively charged) (Fig. 15). This overall energy (fuel) cell situation results in the dendritic membrane of the chemosensory neuron being positively charged at the outer membrane surface in the resting state [82, 83, 84]. Introduction of oxidizing molecules of antifeedant 1,4-naphthoquinones to the outer hydrophilic medium (i.e., in the cuticular pore, pore tubule or sensillum liquor; Figs. 14, 15) withdraws hydrogens from the redox complex in the coupling membrane [83]. This action hyperpolarizes (i.e., increases the protonic potential across) the membrane. If enough molecules of 1,4-naphthoquinone antifeedant withdraw hydrogens from the redox complex associated with the dendritic membrane, then that membrane will become hyperpolarized so that a threshold generator potential is likely to develop less frequently in the involved primary chemosensory neuron [82]. This decreased rate of firing (i.e., impulse generation) in the neuron will result in a temporal "spike" pattern involving fewer action potentials per unit of time (e.g., 5 milliseconds) [82, 85]. Thus, in this chemoreception mechanism of *P. americana* the antifeedant "message" from 1,4-naphthoquinones becomes effectively transferred (transformed) from a molecular-oxidation state in the naphthoquinone, to an increased protonic potential both in the outer aqueous medium bathing the dendrite and outward across the dendritic membrane, to a reduced rate of inward flow of solutes (e.g., Na^+) across the membrane and finally to a reduced rate of action potentials (e.g., chemosensory nerve impulses) which yields fewer impulses per unit of time (i.e., an altered message) passing to the central nervous system (CNS). This continually developing understanding of the energy transduction which occurs during *P. americana* perception of antifeedant 1,4-naphthoquinones allows us to

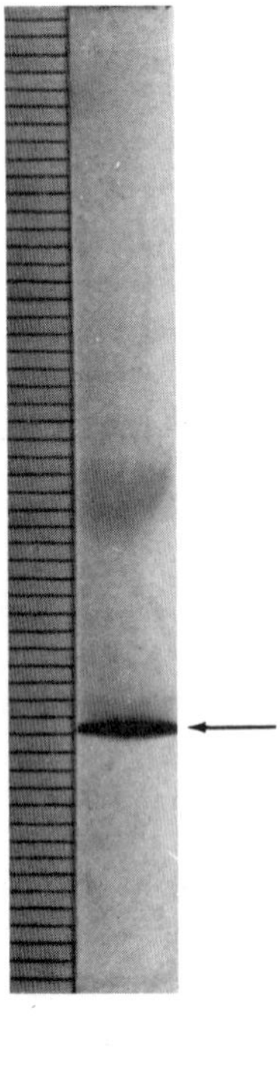

Fig. 16. Single band of redox-complex lipoprotein resolved electrophoretically in 3.5% acrylamide gel after being solubilized from chemosensory nerve membrane with 0.6% Triton X-100 in 0.9% NaCl and purified by affinity chromatography [see [87, 88]]

replace a "black box" of considerable ignorance with a dynamic physiocochemical protonmotive mechanism which consists of explicit, unambiguous structural and functional components. Progress in our knowledge requires a revision of the energy transduction mechanism as proposed earlier [57].

In compliance with the postulates of Michellian energy-coupling theory [81], the three required functional components (Fig. 15) for energy transduction between the dendritic membrane of chemosensory neurons in *P. americana* and antifeedant 1,4-naphthoquinones have been experimentally characterized, and are detailed as follows.

First, the redox complex (Fig. 15) in the dendritic membrane (Fig. 14) is a sulfhydryl-disulfide lipoprotein (Fig. 16). After being solubilized from the nerve membrane with 0.6% Triton X-100 in 0.9% NaCl, this lipoprotein was resolved electrophoretically as a single band in 3.5 or 14% acrylamide gels at pH 7.1 [86]. The lipoprotein was further purified about 57 fold, in terms of μg naphthoquinone bound per μg protein, by affinity chromatography using p-chloromercuricbenzoic acid (PCMB) as a ligand on a 16-carbon sidearm attached to Sepharose 4B [87,88]. The highly purified redox-complex protein subsequently was sharply resolved into three subunits using SDS-disc gel electrophoresis [56]. Estimated molecular weights of the subunits were 19,000; 23,500; and 26,000 Daltons. Amino acid analysis of the redox protein yielded the composition given in Table 5. Content of cysteine sulfhydryls was estimated at $2.8 \pm 0.07 \times 10^{-2}$ M per gram of receptor protein based on bis-[5-carboxy-4-nitrophenyl) disulfide (DTNB) [56].

This sulfhydryl-disulfide protein was shown to bind reversibly with, and reduce electrochemically, antifeedant 1,4-naphthoquinones [3,56,58,59,65,66,68]. The energy-transduction interaction of naphthoquinone with the redox protein produced a fluorescent product, the reduced naphthoquinone [89]. The overall involved energy-transduction interactions between naphthoquinone and the redox protein are given in Fig. 17.

Such sulfhydryl-disulfide proteins make excellent redox-complex components in the energy-transduction machinery of sensory membranes because they are very

Table 5. Amino acid composition a of purified receptor protein from antennae of *Periplaneta americana* adult males

Amino acid	Ratios of Amounts Present	Amino acid	Ratios of Amounts Present
Aspartic acid	1.75	Methionine	0.113
Threonine	0.750	Isoleucine	0.363
Serine	2.178	Leucine	0.967
Glutamic acid	1.885	Tyrosine	0.438
Proline	3.233	Phenylalanine	0.435
Glycine	3.057	Histidine	1.066
Alanine	1.746	Lysine	0.961
Cysteine	Off scale[b]	Ammonia	17.843
Valine	0.542	Arginine	0.590

[a] Analyses conducted in triplicate using a Durrum Model D-500 amino acid analyzer;

[b] Analyses for cysteine sulfhydryls were conducted using the reagent bis-(5-carboxy-4-nitrophenyl) disulfide (DTNB). There were $2.8 \pm 0.07 \times 10^{-2}$ M of detected sulfhydryls per gram of receptor protein (see [65])

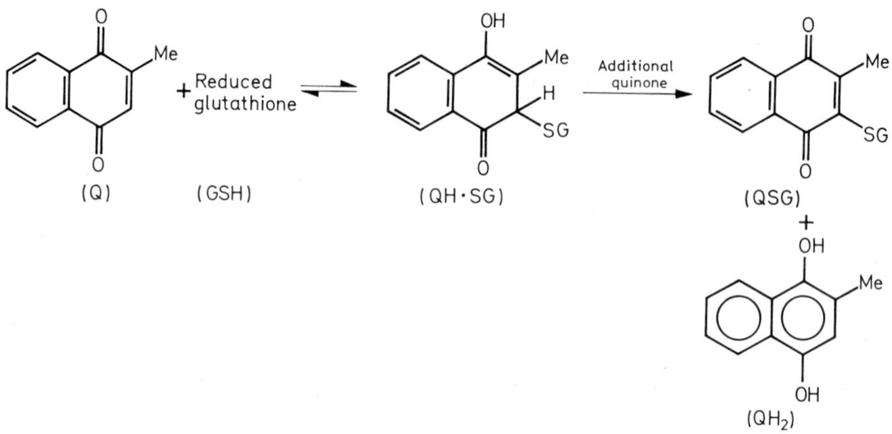

Fig. 17. Primary reaction which occurs between redox-complex receptor protein (reduced glutathione in this model system) and an antifeedant 1,4-naphthoquinone (menadione, i.e., 2-methyl-1,4-naphthoquinone, in this model system) as studied by fluorescence spectrophotometry. *Q* Menadione. *GSH* Reduced glutathione. *QH · SG* Neutral semiradical. *QSG* Oxidized quinone adduct. QH_2 Hydroquinone

sensitive to redox states [78] in the outer or inner aqueous medium (Figs. 14, 15) surrounding the dendritic membrane. As the antifeedant, oxidizing naphthoquinone enters the outer aqueous medium (i.e., contents of the cuticular pore, pore tubule and sensillar lumen) surrounding the dendrite of the chemosensory neuron and becomes reduced (Fig. 17), it will withdraw hydrogens from this medium and from the redox complex in the nerve membrane. Such hydrogen withdrawal thus occurs directly from the redox complex, or indirectly through other contents (e.g., interfacin proteins) in the outer aqueous medium (Figs. 14, 15). This physicochemistry of antifeedant naphthoquinone interactions with the redox-complex (e.g., receptor) protein has been detailed quantitatively [58, 59, 65, 66, 68]. Each antifeedant naphthoquinone at a given molar concentration induces *in vitro* a characteristic quantitative shift in electrochemical (e.g., $E^1/_2$ value) parameters of the redox-complex protein (Table 6). This electrochemical effect (mV shift in $E^1/_2$) on the receptor complex by several naphthoquinones showed the same order of relative values as their relative inhibition

Table 6. Shift in the polarographic half-wave potential (i.e., E 1/2) of each strain (biotype) of receptor in millivolts (mV) as induced by a saturating amount of the indicated 1,4-naphthoquinone

Quinone	Lab[a]		Warf[a]	
	E 1/2	Shift	E 1/2	Shift
Protein alone (control)	−1860 mV	0 mV	−1860 mV	0 mV
2-Methyl-1,4-naphthoquinone	−1875	−15	−1847	+13
1,4-Naphthoquinone	−1881	−21	−1836	+24
5-Hydroxy-1,4-naphthoquinone	−1890	−30	−1827	+33

[a] Biotypes are LAB and WARF

of feeding [58, 64, 66] and electoantennograms [83] or binding to redox-complex protein [75]. A linear relationship was recently demonstrated between any two of these experimental parameters [90]. Simultaneous solution of three resulting equations has allowed the establishment of an unifying mathematical expression for this antifeedant chemoreception process [90]. The excellent agreement among the findings from these several distinct experimental approaches to studying this antifeedant activity strongly supports this SH, SS protein as being the essential redox-complex component [81] in the energy coupling, membrane-associated machinery which enables this *P. americana* perception of naphthoquinones as antifeedants.

A second required membrane component [81] in order for energy transduction to occur between a chemosensory neuron in *P. americana* and antifeedant 1,4-naphthoquinones is a chemiosmotic reversible (Na^+, K^+)-ATPase (Fig. 15). Such a (Na^+, K^+)-ATPase has been detected and characterized in *P. americana* peripheral nerves [84, 91]. The biochemical parameters of this peripheral nerve ATPase were especially documented by the latter workers. The observed range in pH maxima was 8.8–9.5. This enzyme in the presence of 5 mM ATP showed a maximal activity at 5–7 mM Mg^{2+}. The optimal ratio of Mg^{2+} to ATP was 1–1.4. The K_m for Mg^{2+} was 1.2 mM. Increasing concentrations of Na^+ maximized the enzyme activity. The K_m for Na^+ was 6.3 mM, and greatest activity occurred between 90 and 170 mM Na^+. The lowest concentration tested was 10 mM Na^+ because a disodium ATPsalt was used in the assay solution. K^+ was required to hydrolyze ATP, and enzyme activity had a maximal reaction rate in the presence of 20–80 mM K^+. K_m for K^+ was 3.7 mM. The cardiac glycoside ouabain, a highly specific inhibitor of (Na^+, K^+)-ATPase [92], yielded complete inhibition of the *P. americana* peripheral nerve (Na^+, K^+)-ATPase at concentrations greater than 0.1 mM ouabain. Ouabain stimulated this (Na^+, K^+)-ATPase activity at low concentrations near 1×10^{-6} mM. Enzyme activity was almost entirely inhibited by 4.5 mM Ca^{2+}; however, it tolerated lower Ca^{2+} concentrations. Mean data $\pm$ S.E.M. for Ca^{2+} dependence of the (Na^+, K^+)-ATPase were as follows ($n = 5$) (μmoles Pi/mg protein/30 min): 0 mM $CaCl_2$, 2.39 $\pm$ 0.21; 0.1 mM $CaCl_2$, 2.58 $\pm$ 0.27; 0.3 mM $CaCl_2$, 2.58 $\pm$ 0.41; and 4.5 mM $CaCl_2$, 0.33 $\pm$ 0.27. The ATP K_m values for this (Na^+, K^+)-ATPase were 0.24 mM (Lineweaver-Burk plot) versus 0.22 mM (Eadie-Scatchard plot).

The K_m values for Na^+, K^+ and ATP; the Mg: ATP optimum; and the ouabain pI_{50} for *P. americana* peripheral nerve (Na^+, K^+)-ATPase were similar to those reported for a variety of vertebrate and invertebrate (Na^+, K^+)-ATPases. The *P. americana* peripheral nerve (Na^+, K^+)-ATPase has a relatively high range for pH optimum, 8.5–9.5, as compared to those of other (Na^+, K^+)-ATPase systems [93]. It was found to be 7.5 for the excretory system of *Locusta migratoria* [94] and 7.6 for both *P. americana* central nerve cord [95] and larval salivary glands of *Chironomus thummi* [96]. The affinity of the *P. americana* peripheral nerve (Na^+, K^+)-ATPase for K^+ was not as large as that recorded for other (Na^+, K^+)-ATPase. This peripheral nerve (Na^+, K^+)-ATPase did not have single optimal concentrations for Na^+ and K^+. This may reflect the large local changes in ionic concentrations, especially of Na^+, which occur in nerve membranes during the conduction of action potentials. Insect (Na^+, K^+)-ATPases tend to be slightly more sensitive to ouabain than do such vertebrate enzymes [93]. The observed ouabain pI_{50} of 6.2 for the *P. americana* peripheral nerve enzyme was in close agreement with other such values from invertebrate systems [95, 96]. The observed slight stimulation of the peripheral nerve (Na^+, K^+)-ATPase activity by low (1×10^{-9} M) ouabain has been noted with enzyme from other tissues [93, 96, 97]. The *P. americana* peripheral nerve (Na^+, K^+)-ATPase does not appear to be Ca^{2+} dependent. Inhibition of other (Na^+, K^+)-ATPases by Ca^{2+} however has been reported [98, 99].

The *P. americana* peripheral nerve (Na^+, K^+)-ATPase demonstrated the properties common to the enzymatic component of sodium pumps (i.e., solute porter systems, [81]) characterized from other tissues: (1) location in the plasma membrane, (2) requirement of both Na^+ and K^+ for activity and (3) inhibition by ouabain [92, 93, 100, 101]. The observed cation K_m values for *P. americana* peripheral nerve (Na^+, K^+)-ATPase are similar to those of other (Na^+, K^+)-ATPase shown to be integral parts of sodium pumps [93]. Thus, the (Na^+, K^+)-ATPase characterized from *P. americana* peripheral neurons demonstrates the capabilities and functions in the dendritic membrane which are required [81] for it to participate in the energy transduction between antifeedant 1,4-naphthoquinones and the involved chemosensory neuron.

Thirdly and lastly, the energy transduction requires a solute (e.g., Na^+) porter system (i.e., pump) [81]. Our studies of the *P. americana* peripheral nerve (Na^+, K^+)-ATPase have indicated that such a pump (i.e., porter system, [81]) is intimately associated functionally with the enzyme [57, 102]. Solute porter systems, like (Na^+, K^+)-ATPases, are typically located in plasma membranes such as dendritic membranes, require Na^+ and K^+ for activity and are inhibited by ouabain [80, 81, 93]. These requirements were demonstrated in the *P. americana* peripheral nerve system [57, 91].

8.1.3 Electrophysiological Parameters

Electrophysiological investigations have shown that some allomones (i.e., antifeedants, repellents, deterrents) elicit a readily discriminable behavior in the organism which is relatable to the impulse pattern which that chemical stimulates in exposed primary (i.e., afferent) chemosensory neurons [82]. In such situations the behavioral responses of the organism may be limited by the descriminative powers of its peripheral sensory neurons. If so, two compounds may be discriminated behaviorally in such an organism if they elicit some form of differential response either within or across the relevant primary sensory neurons [82]. In such cases encoding of chemical information must occur in the primary sensory neurons because different behavioral responses occur. Working with such afferent-efferent systems, the rules (codes) which govern the mapping of the input variable onto the output variable should be decipherable [103]. Unfortunately, there are only a very few insect species from which sufficient behavioral and electrophysiological information for a common set of compounds is available [82].

The following specific examples seemingly indicate some of the minimal requirements which must be met before one can form conclusions about the encoding process in such chemoreception. The presented data were collected electrophysiologically from whole antennae via electroantennograms (EAG's) (Fig. 4) or from a single sensory hair (sensilla). Many different indices of neutral activity have been applied to the recorded electrophysiological responses of tested chemosensory neurons [82]. There is nothing inherently correct about any particular index of such activity. The simplest rules which allow the prediction of the behavioral outflow, knowing the electrophysiological input via primary afferent chemosensory neurons, should be adequate [82].

There is a large quantity of behavioral evidence which demonstrates that the blowfly adult (*P. regina*) can discriminate among a large number of different chemicals based on electrophysiological input by a single sensillum containing not more than four chemoreceptor neurons [60]. Known antifeedant compounds for this fly include a variety of salts, acids, alkaloids and alcohols. Neural activity in the so-called "salt" receptor neurons of the single sensillum shows a good correlation with elicited behavioral rejection (i.e., antifeedant actions).

Adult female *Aedes aegypti* mosquitoes possess a distinct grooved-peg, type A3, sensillum on their antennae. Each sensillum contains three receptor neurons which are especially sensitive to some components in human sweat, e.g., lactic acid, and to the repellent, DEET, which is *N*, *N*-diethyl-*m*-toluamide [85]. DEET and other compounds known to inhibit female mosquito

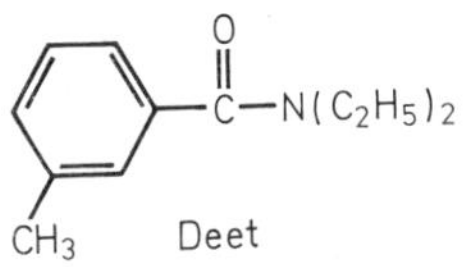

Deet (20)

response to lactic acid inhibit all three neurons in the type A3 sensillum. DEET reduced the firing rate of each neuron type whether it was applied alone or in combination with lactic acid [85]. Thus, an inhibitory compound may exert its effect in this system by inhibiting the firing rates of the same set of receptor neurons which are responsive to excitants (i.e., lactic acid).

A prior treatment with a antifeedant 1,4-naphthoquinone reduces the millivolts of depolarization (Fig. 4) induced in an EAG of *P. americana* by a standard dose of the excitant amyl acetate [83]. The order of the relative EAG inhibitions by the tested 1,4-naphthoquinones matched the order of their relative inhibitions of feeding and order of combined redox potentials and hydrogen-bonding capabilities [3,57,90]. In this *P. americana* chemoreception system, depolarization of the receptor potential beyond a threshold level excites (fires) those nerves; whereas, hyperpolarization inhibits such impulse activity. Antifeedant 1,4-naphthoquinones normally hyperpolarize the dendritic membrane of, and thus decrease the firing (impulse) rate of, primary *P. americana* chemosensory neurons; and through this altered electrophysiology such molecules elicit an inhibition of feeding.

The limited electrophysiological evidence regarding nerve-impulse activity correlated with antifeedant versus feedant behavioral activities in the above insect examples indicates that compounds elicit antifeedant actions through decreasing the firing rate of pertinent primary chemosensory neurons. In the case of 1,4-naphthoquinones, it is established that those tested at a given concentration produced an order of relative decreases in the firing rate of pertinent primary chemosensory neurons which matches the order of their relative inhibitions of *P. americana* feeding. Such pertinent neurons may, or may not, also respond to excitants.

8.2 Pharmacological Alterations of Antifeedant Responses

Most successful efforts at altering organismal antifeedant responses to given stimulants, via the application of reagents and pharmacologically active agents (drugs), have been directed topically at the pertinent sensory organs and neurons [3,57,71]. Such chemical modification has been utilized in behavioral, electrophysicological, biochemical and biophysical analyses of sensillar- and membrane-related phenomena in gustatory receptor functions. Special emphasis has been placed on using chemicals which have a known mechanism of action [71].

8.2.1 Alterations through Proteins

Because proteins play essential roles both in gustatory sensory organs and in membranes of primary chemosensory neurons [57,58,64,65,67,71], many specific reagents for proteins have been employed in these studies. Such specific reagents may interact with both saline-soluble or membrane-associated proteins.

Sulfhydryl proteins have proven especially important as components of chemoreception mechanisms both at the level of the sensillar liquor (i.e., fluid) (Fig. 14) and sensory nerve membrane [3,56,57,58,59,65,66,67,69,71,75,104]. Sulfhydryl (SH) groups are highly susceptible to oxidation, substitution, addition and metal-binding reactions under physiological conditions [105,106]. Thus, many reagents and drugs can be used to detect such functional groups. However, the specific reactivity of a particular sulfhydryl (thiol), or other functional group, is greatly influenced by the local protein environment. Thiol groups especially may be in a "buried" state, i.e., sterically hindered by other protein structure. Hydrogen bonding, charge relationships and hydrophilic versus hydrophobic interactions are other environmental variables which may influence the reactivity of such groups. Thus, a given sulfhydryl reagent only reacts with a portion of the sulfhydryls present. It is especially due to these facts and conditions that the contents of cuticular pores, pore tubules and sensillum liquor are so important in determining thresholds of antifeedant sensitivity for specific primary chemosensory neurons [3,56,57,66,107]. These contents, especially because of their sulfhydryl-disulfide *interfacin* proteins, readily bind various antifeedant molecules (e.g., 1,4-naphthoquinones), free radicals and ions and thus may significantly alter the threshold number of such antifeedant-stimulant units (e.g., molecules) required in a chemosensory sensillum to alter the firing rate of the primary afferent neuron such that antifeedant behavior results [3]. In some messenger situations, the fluid contents of pores and cavities in the chemosensory structure thus may function as a "chemical sink" to buffer the primary chemosensory nerve from the external chemical environment [3]. In other messenger circumstances, those contents may serve to amplify the energy exchange between introduced antifeedant molecules and the dendrite of the primary chemosensory neuron [3,56,57,65,66,107]. These vital "yin-yang" capabilities [3,57] of the contents of cuticular pores, pore tubules and sensillum liquor are especially attributable to the above described (Fig. 13) saline-soluble, sulfhydryl-containing *interfacin* proteins.

Demonstrated differences [90] in the thiol content, and thus in the antifeedant-binding capabilities, of interfacin proteins in two genetically distinct populations of *P. americana* have been shown to contribute significantly to their biotypically distinguishable sensitivities to 1,4-naphthoquinone antifeedants [3,56,57,66,68].

Alkylating reagents are among the more useful chemicals for detecting functional sulfhydryls in proteins. N-Ethylmaleimide (NEM) is such a reagent which not only reacts with SH's, as follows,

$$R_1{-}SH + R_2{-}CH{=}CH{-}R_3 \rightarrow R_1{-}S{-}\underset{\displaystyle \diagdown CH_2{-}R_3}{\overset{\displaystyle \diagup R_2}{CH}} \tag{21}$$

but in doing this it fully mimics 1,4-naphthoquinones as antifeedants to *P. americana* [3, 56, 57, 58, 65, 66, 107]. NEM at 10^{-2} M, applied to the antennae of *P. americana*, also sensitized such insects to subsequent exposures to the antifeedant 1,4-naphthoquinone, juglone (1.25×10^{-5} M) [65]. Such a sensitization can be caused by NEM binding to SH's in interfacin proteins and thus reducing their abilities to buffer chemosensory nerves subsequently against the 1,4-naphthoquinone, juglone [3]. The relative effects of N-substituted maleimides on gustatory receptor systems correlate well with the ether: water partition coefficient of each compound [71].

Other SH reagents which react through alkylation, but not by addition to double bonds, also affect gustatory responses [71, 83, 104, 108]. Iodoacetic acid (IAA) and iodoacetate are examples of such reagents. It was shown that IAA and the 1,4-naphthoquinone antifeedants bind competitively to sulfhydryl-rich sites in protein from sensory structures [75].

Organo-mercurials (OM's) react with sulfhydryls by a reversible mercaptide-forming mechanism, as follows

$$R_1 - SH + Hg - R_2 = R_1 - S - Hg - R_2 \tag{22}$$

In the presence of excess thiol, an exchange reaction may occur as follows

$$R_1 - SH - Hg - R_2 + R_3 = R_1 - SH + R_3 - S - Hg - R_2 \tag{23}$$

OM's have proven to be much more effective than alkylating reagents in modifying gustatory receptor cells in animals [71]. p-Chloromercuric benzoate (PCMB) was twenty times as effective as NEM in depressing the sugar taste response in the rat [108]. PCMB is also more inhibitory than NEM to the sugar receptor in insect sensilla [71]. The differences in effects of PCMB versus NEM apparently involve their different mechanisms of reaction with thiols in proteins. They also are likely to react with different SH's within a given protein, and thus to have distinct effects on that protein.

It is now clear that oxidation of sulfhydryls in a redox-complex protein in the dendritic membrane is involved in the energy transduction associated with the perception of, at least, some antifeedants [3, 57]. Thiols may be oxidized via several mechanisms [78]. Oxidizing agents show varying degrees of specificity for sulfhydryl groups [105]. Thiols also vary greatly in their ease of being oxidized depending on the adjacent amino acid sequence, and various parameters of the environment surrounding the protein. These variable characteristics among sulfhydryls within, and among, proteins may provide bases for specificity of interactions with feeding inhibitors and excitants.

Reducing agents, such as dithiothreitol (DTT), reduce disulfide bonds. Such bonds are important in stabilizing the tertiary structure of proteins. Through such actions, reducing agents may be used to alter the activities of chemoreception cells [71]. DTT at 10 mM for 20 minutes on taste sensilla of fleshfly yielded a reversible reduction in receptor response to sucrose [109]. Perfusing DTT onto silkmoth antennae however slightly increased the EAG response magnitude to sex pheromone [104]. As an antennal treatment, DTT at 10^{-3} or 10^{-2} M increased the behavioral acuity of two *P. americana* biotypes to juglone [65, 66]. The weaker reducing agent, reduced glutathione (GSH), at 10^{-2} M, had opposite effects on the behavioral responses of the two *P. americana* biotypes (i.e., WARF and LAB) to the antifeedant juglone [65, 66]. The treatment sensitized the naturally less-sensitive WARF strain (biotype) but desensitized the naturally more-sensitive LAB strain. Other studies had already shown that the interfacin proteins and receptor mechanisms of these two biotypes of *P. americana* differed in their sulfur chemistry and other parameters [90]. Thus, the opposite effects of GSH in this treatment just further confirmed this difference.

Amino-group reagents have shown some usefulness in altering insect feeding responses when used under alkaline (e.g., pH 8.0) conditions [110]. 2,4,6-Trinitrobenzene sulfonic acid (TNBS), at 2 mM for 2–3 minutes on fleshfly labellar hairs, almost completely inhibited response to fructose.

Some proteases have been used effectively to alter insect sensory responses to taste stimulants [71]. Pronase was the most effective among those tested.

8.2.2 Alterations through Lipids

In addition to proteins, various heterogeneous lipids are important components (a) of the contents of the cuticular pores, pore tubules and other chemosensillar cavities; and (b) of course the dendritic membrane. Thus, reagents which alter these various lipids may be useful in pharmacologically manipulating insect responses to antifeedants. Lipids have been theorized as being especially important in those membrane phenomena involved in chemoreception of sour, salt or bitter stimuli [71, 111]. Many bitter-tasting antifeedants are lipophilic substances and thus may interact especially with the lipid components of dendritic membranes [71].

The surface potential and permeability properties of the dendritic membrane may be significantly altered by interference with the complexing of Ca^{2+} with anionic sites in membrane phosphilipids. Di- and tri-valent cations of a large ionic radius may be especially effective in such competition with Ca^{2+} [112]. These polyvalent cations may act as stabilizers of the membrane potential through reducing the magnitude of membrane-conductance change during depolarization. Local anaesthetics also affect nerve membrane functions by this means [113]. Local anaesthetics and amphiphilic bitter antifeedants also have a similar initial action on chemoreceptor cells. The effect apparently is an initial membrane depolarization which is associated with an increased membrane-input resistance [114]. This similar action has been attributed to the lipophilic nature of both sets of compounds.

8.2.3 Summary Comments

In conclusion, pharmacological agents (drugs) and reagents have been used effectively to alter insect and other animal responses to antifeedants [57, 71]. Most successful treatments have been directed at sulfhydryl-disulfide proteins and various lipids in the chemoreception mechanism.

9 An Energy Exchange Basis for a Chemical Ecology Code

It is now well established that messengers in chemical ecology function, at least, as "double agents". They are allomonic to some species or biotypes, but kairomonic to others. A chemical messenger thus tells some species or biotypes that they are in a favorable (i.e., kairomonic) environment while simultaneously telling others that they are in a hostile (i.e. allomonic) habitat. Classical examples of this basic chemical ecological principle include the following: (1) The dimeric sesquiterpene gossypol and related compounds in cotton plants are kairomonic to the cotton boll weevil (*Anthonomus grandis*), but are allomonic (e.g., inhibitory to feeding) in cotton plants to *Heliothis* spp. and *Epicauta* sp. [9,10]; and (2) 5-hydroxy-1,4-naphthoquinone (juglone) in *Carya* trees is kairomonic to *S. quadrispinosus* (the hickory bark beetle), but is allomonic (e.g., inhibitory to feeding) in *Carya* to *S. multistriatus* (the smaller European elm bark beetle) that naturally feeds on elms (*Ulmus*) [7,8].

Such a dual function for each messenger in chemical ecology provides a binary energy-based (e.g., millivolt-based) mechanism for ordering the distributions and population densities of species or biotypes among, and within, ecological habitats [2]. This occurs because a species is likely to occur where kairomones for it are dominant and to be scarce, or absent, where allomones for it are dominant. Because each chemical messenger is, at least, allomonic to some species or biotypes and kairomonic to others, populations of several species will likely overlap in a highly ordered fashion in and among habitats.

The above described binary energy-based mechanism seemingly provides a functional basis for an energy exchange code (EEC) in chemical ecology. Such a code must simply constitute a set of rules which governs the mapping of an input variable onto an output variable [82,103]. In the specific situation of *P. americana* and 1,4-naphthoquinones, the input variable is the amount (i.e., moles) of the chemical messenger (i.e., $(X)M_{99}$, Table 7) required in the ecological environment (i.e., behavioral assay chamber) surrounding the insect. In this case the primary output variable (i.e., maximum mV E1/2 shift in vitro by the redox-complex protein; mV Max, Table 7) is linearly related (Fig. 18) to the messenger input variable, $(X)M_{99}$. Thus, the requirements for a code readily exist in this biological system. Though the primary output, mV Max, by the redox-complex protein for each naphthoquinone is quantitatively similar ($P < 0.01$) in both biotypes of *P. americana* (Fig. 18), it is opposite in charge (i.e., $-$ versus $+$), and thus is electrochemically distinctive, between the two (i.e., LAB versus WARF) phenotypes. Additionally, the primary output variable, mV Max, shows a firm relationship with a secondary quantitative neurophysiological output component, i.e. % mV inhibition of the electroantennogram (EAG) elicited by the standard excitant amyl acetate [90]. A 5 mV shift in E1/2 equals an 8% mV inhibition of the EAG. The above experimentally determined facts allow the involved mV-based, energy-exchange code (EEC) in chemical ecology to be expressed mathematically as $\text{Log } Y = 3.40 - 0.112 (\text{Log } X)$, and graphically in Fig. 19. In this equation, "Y" equals the primary output (efferent) variable, mV Max (Table 7), and "X" equals the input (afferent) variable, $(X)M_{99}$ (Table 7). Changes in mV Max dictate the resultant behavior of the whole insect. This code thus enables the organism (i.e., *P. americana*) to determine neurophysiologically (i.e., afferently) whether it is in an environment which is chemically hostile (i.e., allomonic), and also to direct neuro-

Table 7. Numerical values for the moles of specific 1,4-naphthoquinones required to yield a >99% avoidance behavior in each of two biotypes of *P. americana* and for the maximum elicited *in vitro* E 1/2 shift in millivolts by the redox-complex protein of each biotype in response to a saturating amount of a 1,4-naphthoquinone

Chemical messenger[d]	$[X]M_{99}$[a]		mV Max[b]	
	LAB[c]	WARF[c]	LAB	WARF
5-OH	1	5	−30	+33
1,4-	25	40	−21	+24
$2\text{-}CH_3$	500	1000	−15	+13

[a] $[X]M_{99}$ = Times (X) of moles of the given messenger required for >99% *P. americana* avoidance behavior in comparison with that quantity (i.e., M_{99}) for 5-OH and the LAB biotype being set at one (1). One (1) equals 2×10^{-5} Moles;

[b] mV Max = Maximum *in vitro* polarographic E 1/2 shift in mV by the redox (receptor) protein for the indicated *P. americana* biotype elicited by a saturating concentration of the indicated 1,4-naphthoquinone antifeedant;

[c] Behaviorally distinct biotypes of adult *P. americana* with regard to perception of 1,4-naphthoquinones as allomones (e.g., repellents and antifeedants);

[d] 5-OH represents 5-hydroxy-1,4-naphthoquinone; 1,4-, 1,4-naphthoquinone; and $2\text{-}CH_3$, 2-methyl-1,4-naphthoquinone

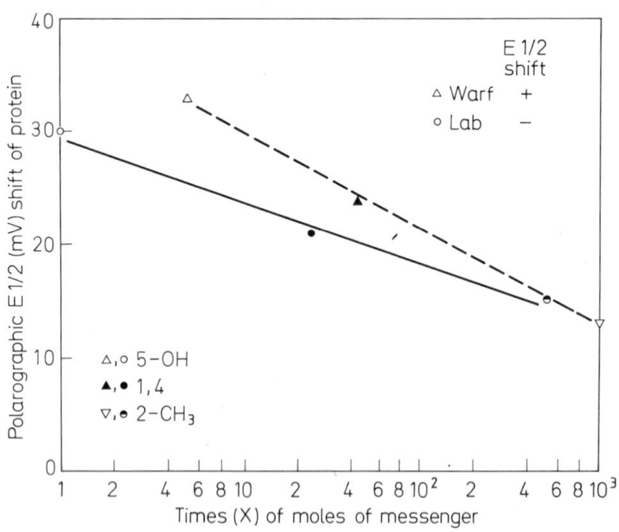

Fig. 18. Linear relationships between the moles of each indicated 1,4-naphthoquinone required to behaviorally deter >99% of the assay insects of each behavioral biotype, i.e. LAB vs. WARF, of *P. americana*; and the maximum millivolt (mV) shift in the polarographic E 1/2 value of each biotype of redox-complex (receptor) protein induced *in vitro* by saturating amounts of each antifeedant

physiologically (i.e., efferently) the actions of those muscles required for it to express an appropriate (e.g., avoidance) behavior. These experimental findings demonstrate that regarding *P. americana* adults and their perception of 1,4-naphthoquinones, the peripheral chemosensory system involved in the sensilla (hairs) of these insects can encode biophysically all the information required to the insect's nervous system to enable its decision to avoid behaviorally the environment treated with these allomonic (e.g., antifeedant) chemicals. Beyond connecting afferent and efferent neurons, the central nervous system (CNS) of *P. americana* may be involved in some unknown manner in this chemosensory-avoidance behavior event, but our known facts negate such CNS activity as a necessity with regard to the observed event. It is now evident that the neurophysiological encoding of a message which elicits the observed avoidance behavior in *P. americana* entails the reduction of the impulse rate in affected chemosensory neurons below a critical threshold rate. Our overall investigations with 2-methyl-1,4-naphthoquinone (menadione) as an antifeedant for *P. americana* document that an inhibition as small as 8% in a standard amyl acetate-excited EAG switches the resultant behavior from excitation (+) to inhibition (—) [90].

This finding of how an abiotic (chemical) parameter in the external environment of a higher organism (i.e., insect) can interact bioenergetically with its sensory system to elicit a genetically programmed phenotypic behavior constitutes a new understanding of how energy is exchanged as information between abiotic and biotic environments.

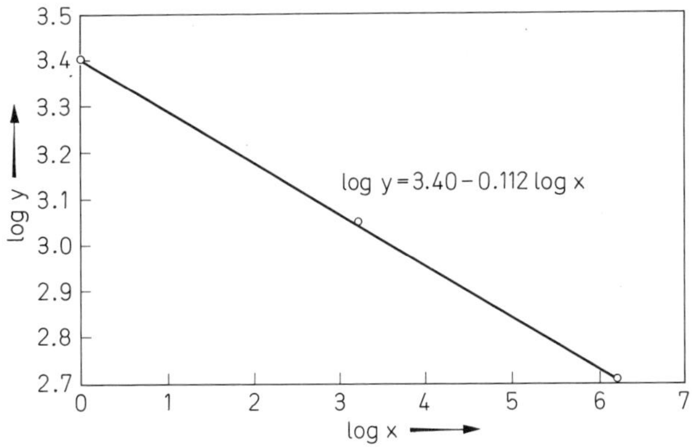

Fig. 19. Linear relationship which exists between the log of the primary input (afferent) variable, X (i.e., $(X)M_{99}$, Table 7), and the log of the primary output (efferent) variable, Y (i.e., mV Max, Table 7), in *Periplaneta americana* perception of 1,4-naphthoquinones whereby changes in mV Max dictate the resultant behavior of the whole insect. Relationship thus describes the quantitative conversion of an abiotic energy state in repellent 1,4-naphthoquinones into a dictated behavior (i.e., avoidance) by a whole animal

10 Summary and Conclusions

When insects are prevented from, or inhibited in, feeding, the results for the insect may range from slower growth, to less growth, to reduced reproduction, to reduced longevity including sudden death. Coevolution between insects and hosts has involved allomones, including antifeedants, as major parameters of the relationships. Host defense thus is based significantly on such compounds. Man should be able to benefit by further attempting to exploit, augment or amplify natural situations which involve antifeedants.

The effects of insect feeding on hosts, whether plant or animal, may be the difference between sustained host life and sudden death. From more practical agronomic or husbandry perspectives, reduction of insect feeding may make the difference between a harvestable, economically profitable crop or herd, and disaster. In terms of reducing human discomfort and disease, antifeedants (including repellents) may play paramount roles in our daily lives. This may be especially true in semi-tropical and tropical environments where insect-transmitted pathogens of humans may be more prevalent. Thus, expanded understanding of the antifeedant actions of chemicals should enable their more widespread effective use to suppress insect feeding in a broad range of pest situations.

10.1 General Understanding of Antifeedant Action

Knowledge on chemicals which have antifeedant activities is growing quite rapidly, but known structures of such compounds, exclusive of repellents, still may be numbered in the dozens. Our awareness of phytochemicals which have antifeedant actions is especially growing rapidly. Much research on defensive secretions of insects also has contributed to the understanding of antifeedant activities of chemicals. From these total studies, it is becoming increasingly obvious that plants and animals frequently utilize similar, if not identical, compounds as antifeedants against predators and parasites.

Though some chemical structures (e.g., oxidants) have been more frequently associated with allomonic than kairomonic activity, our understanding strongly suggests that one should only label a compound as an antifeedant if one also states the species of affected insect and the specific test conditions. This is recommended because one insect species' antifeedant is another insect species' feedant. Each chemical messenger in chemical ecology seems to be, at least, a "double agent". It is kairomonic to some species or biotypes, and allomonic to others. Such an emerging understanding of chemical ecology indicates that an energy-exchange (transfer) code (EEC) may (must) function among individuals, within and among species and biotypes, and their environments.

10.2 Deciphering an Energy-Exchange Code for Antifeedant Activity

An energy-exchange code (EEC) apparently exists in the chemical ecology of antifeedant activity. Current understanding of the antifeedant activities of 1,4-naphthoquinones and 1,4-benzoquinones against certain insects strongly supports the hypothesis that such a code exists. For the first time, the effects of an environmental chemical (i.e., quinone) on the behavior of an animal (phenotype) via its nervous system are explainable mathematically in millivolt (mV) units at the messenger quinone, energy-transducing receptor, dendritic membrane-polarization and nerve-impulse levels of the involved afferent-efferent system. Linear quantitative relationships exist between all of these parameters of the system; thus, any one can be expressed directly in terms of any other parameter [90]. Because knowledge of the input component (i.e., moles of a specific quinone) to the primary neuron in the chemosensory system of a known biotype of an insect species (*P. americana*) allows the direct quantitative prediction of an output component (i.e., mV shift in E 1/2 value of the redox-complex protein), an adequate basis for an informational code exists and it can be expressed quantitatively in millivolts (mV) of energy. Even though the central nervous system (CNS) may play additional roles in this described afferent-efferent process associated with the antifeedant activity of quinones, the now proven information-encoding capabilities of the *P. americana* peripheral nervous system (PNS) indicate that the CNS only must connect the essential afferent and efferent nerves. Related investigation with these same quinones and *S. multistriatus* beetles confirms the much more detailed findings from the studies with *P. americana*. The current understanding of the energy-exchange code (EEC) between quinones and *P. americana* indicates that such messengers inhibit, or prevent, feeding through decreasing the firing (i.e., impulse) frequency of exposed chemosensory neurons below a definable threshold level. Findings from those few other electrophysiological-behavioral investigations of antifeedant effects in insects seem to support inhibition of chemosensory neuron-impulse frequency as the basis for the information transfer (encoding). Chemically speaking messenger quinones oxidize (i.e., withdraw hydrogens from) the redox-complex, energy-transducing lipoprotein receptor in the dendritic membrane of the insect chemosensory neuron, this hyperpolarizes the membrane and the net result neurophysiologically is a reduced impulse frequency which tells the animal to reduce, or stop, feeding.

Acknowledgements. Our research on chemoreception in insects was supported by the College of Agricultural and Life Sciences, University of Wisconsin, Madison; and in part by research grants No. GB-6580, GB-8756, GB-41868 and BNS 74-00953 from the U.S. National Science Foundation. My thanks are extended to the many coworkers who have helped along the way. Special thanks are given to Mary Dietrich and Kim Viney for invaluable technical assistance.

11 Glossary of Special Terms

Action potential:	The potential perpetuated in a neuronal membrane when the cell body receives a threshold generator potential.
Afferent:	Sensory neural functions.
Allelochemical:	A toxic chemical released by one species which affects another species.
Allomone:	A molecule, ion or free radical which is operative as an interspecific messenger in chemical ecology, and is adaptively advantageous to the emitter (releaser) of the substance.
Antifeedant activities:	Treatment effects which reduce or prevent feeding.
Antifeedant:	A substance which prevents, or reduces, feeding.
Assay substrate:	The material on which, or in which, the test substance is presented in the bioassay.
Behavioral biotype:	A behaviorally distinguishable population of individuals within a species.
Binary:	Consisting of two parts.
Binding site:	Discrete region in a macromolecule (e.g., receptor) where it complexes with given ligands.
Bioenergetics:	The biology of energy conversions.
"Black box":	A label frequently applied to a realm of biology which is not understood.
Chemical profile:	Species of chemicals contained.
Chemical sink:	An identified repository for one or more species of molecule, ion or free radical within a given chemical environment.
Choice test:	Bioassay in which the insect is given a choice between two or more treatments.
Compartmentalization:	Non-random distribution of an entity (e.g., stimulant among cells of a tissue).
Cultivar:	Variety of plant.
Cytotoxin:	A chemical which is toxic to cells.
Decompartmentalization:	Distribution of a biochemical beyond the cells where it naturally occurs.
Defensive chemical:	Compound which elicits an avoidance behavior by the perceiver.
Depolarization:	Reduction in the level of charge distributed across a membrane between inside (—) to outside (+) aqueous media.
Deterrent:	A chemical which discourages (deters) a given organismal behavior (e.g., feeding).

Double agent:	A chemical which is allomonic to some organisms and kairomonic to others.
Drinking:	Ingestion of a liquid.
Efferent:	Motor neural functions.
Electrochemistry:	The chemistry associated with electricity.
Electromotive force:	That difference of electronic potential which causes a flow of current.
Electrophilic substance:	An electron-seeking entity.
Encoding process:	The bioenergetic mechanism by which energy is exchanged between messenger ligands and neuronal receptors such that information is created.
Energy exchange:	Synonymous with energy transfer.
Energy transduction:	The conversion of energy from one state to another.
Energy transfer:	Exchange of energy (e.g., hydrogens) between two or more entities.
Excitant:	A stimulant which increases the level of a given behavior by an organism.
Feeding response:	Increase or decrease in feeding in a bioassay, attributable to treatment.
Feeding:	The act of eating.
Firing rate:	The number of action potentials generated per unit of time (e.g., 100 milliseconds).
Generalist species:	A species that practices polyphagy.
Hematophagy:	Feeding on blood.
Herbivore:	An animal that eats plant tissue.
Host range:	Grouping of species or varieties which an insect will use for food or shelter.
Host:	A live organism that serves as a food substrate or shelter.
Impulse activity:	Action potentials generated by a neuron in a given time interval.
Impulse generation:	A neuron generates an action potential.
Induced wounding reaction:	A reaction associated with wounding which is induced by environmental parameters.
Inducible chemical:	A chemical produced in an organism (e.g., plant) when it is placed under environmental stress.
Ingestion:	To take a substance into the alimentary tract.
Inhibition:	Reduction of behavioral activity level (e.g., feeding).
Input variable:	The changeable input occurring in some input: output relationship.

Interspecific messenger:	A chemical which serves as a messenger among two or more species of organisms.
Intraspecific situation:	One which occurs among individuals within a species.
Kairomone:	A molecule, ion or free radical which is operative as an interspecific messenger in chemical ecology, and is adaptively advantageous to the perceiving organism.
Messenger:	A chemical which passes between individual organisms and interacts with receiving individuals in ways which convey information.
Michael adduct:	The result of a nucleophile undergoing conjugate addition to an activated double bond.
Monophagy:	Feeding on one species, or a very few species, in one genus.
Neutral substrate:	Substrate which does not influence feeding in a bioassay.
No-choice test:	Bioassay in which the insect is exposed to only one candidate substrate for feeding.
Non-host range:	Grouping of species or varieties which an insect will not use for food or shelter.
Non-host:	A live organism that is not serving as a food substrate or shelter for the insect of interest.
Nucleophilic terminus:	A chemical entity which supplies electrons to a electron deficient center.
Oligophagy:	Feeding on several species in a few genera in one or a very few families.
Output variable:	The changeable output occurring in some input: output relationship.
Perception:	Direct acquaintance (recognition) with anything through the senses.
Phytoalexins:	Chemicals produced in plants under a given stress (e.g., a pathogen) which increase such plants' resistance to those stresses.
Polyphagy:	Feeding on species in numerous genera in several families.
Positive control:	A known treatment which elicits the desired behavior (e.g., feeding).
Prey:	A host that was killed by predator eating.
Primary afferent neuron:	The most peripheral sensory neuron in a neural circuit.

Primary output:	The first form (e.g., level) of output in some input: output relationship.
Protein subunits:	The sequences of amino acids which combine to form a native protein.
Proteinease inhibitor:	A chemical which inhibits the enzymatic activity of proteineases.
Protonmotive force:	That difference of protonic potential which causes flow of current.
Receptor neuron:	The nerve cell which contains the pertinent receptor.
Receptor:	Macromolecular entity which binds (complexes with) a chemical, ion or free radical ligand (messenger).
Redox protein:	A protein which is sensitive to oxidative-reductive changes in its environment.
Reducing agent:	The entity responsible for a reducing reaction.
Repellent:	A volatile substance which elicits an avoidance behavior by an organism (e.g., insect).
Reversible reduction:	A reducing reaction where re-oxidation may subsequently occur.
Solute porter system:	A molecular entity (mechanism) in a membrane which functions to pump or transport cations or anions across the membrane against their respective gradients.
Specialist species:	A species that practices monophagy.
Specific feeding excitant:	A known chemical, ion or free radical which excites feeding.
Stimulant:	A chemical which alters the behavior of an organism through its nervous system.
Substrate:	Any substance ingested in conjunction with the act of feeding.
Systemic signal chemical:	A chemical (e.g., PIIF) which moves systemically in an organism (e.g., plant) and functions as a Messenger.
Temporal spike pattern:	The pattern of action potentials generated by one or more neurons within a given time interval.
Threshold generator potential:	The potential required in the dendritic membrane in order for the neuron (cell body) to fire, i.e., generate an action potential.
Treated positive control:	Positive control plus candidate substance presented in, or on, the same bioassay substrate.

12 References

1. Brown, W. L., Jr., Eisner, T., Whittaker, R. H.: Bioscience *20*, 21–22 (1970)
2. Norris, D. M., Kogan, M.: Biochemical and morphological bases of resistance. In: Breeding plants resistant to insects, pp. 23–61. Maxwell, F. G., Jennings, P. R. (ed.). New York: John Wiley and Sons 1980
3. Norris, D. M.: Bull. Entomol. Soc. Amer. *22*, 27–30 (1976)
4. Klun, J. A., Tipton, C. L. Brindley, T. A.: J. Econ. Entomol. *60*, 1529–1533 (1967)
5. Ikeda, T., Matsumura, F., Benjamin, D. M.: Science *197*, 497–499 (1977)
6. Jermy, T.: The host-plant in relation to insect behavior and reproduction. New York: Plenum Press 1976
7. Gilbert, B. L., Baker, J. E., Norris, D. M.: J. Insect Physiol. *13*, 1453–1459 (1967)
8. Gilbert, B. L., Norris, D. M.: J. Insect Physiol. *14*, 1063–1068 (1968)
9. Maxwell, F. G., Lefever, H. N., Jenkins, J. N.: J. Econ. Entomol. *58*, 792–793 (1965)
10. Maxwell, F. G. et al.: Entomol. Exp. Appl. *12*, 801–810 (1969)
11. Lukefahr, M. J., Houghtaling, J. E.: J. Econ. Entomol. *62*, 588–591 (1969)
12. Kogan, M.: Proc. Internat. Congr. Entomol. *15*, 211–248 (1977)
13. Kubo, I. et al.: J. Chem. Soc. Chem. Comm. *1013* (1976)
14. Kubo, I., Nakanishi, K.: Insect antifeedants and repellents from African plants. In: Host plant resistance to pests, pp. 165–178. Hedin, P. A. (ed.). Washington, D. C.: Am. Chem. Soc. 1977
15. Dethier, V. G.: Chemical interactions between plants and insects. In: Chemical ecology, pp. 83 to 102. Sondheimer, E., Simone, J. B. (ed.). New York: Academic Press 1970
16. Swain, T.: Proc. Internat. Congr. Entomol. *15*, 249–256 (1977)
17. Chiang, H.-S., Norris, D. M.: J. Agr. Food Chem. *31*, 726–730 (1983)
18. Sturchkow, B., Low, I.: Ent. Exp. Appl. *4*, 133–142 (1961)
19. Schreiber, K.: Ent. Exp. Appl. *1*, 28–37 (1958)
20. Dahlman, D., Hibbs, E. T.: Ann. Entomol. Soc. Amer. *60*, 732–740 (1967)
21. Schoonhoven, L. M.: Secondary plant substances and insects. In: Recent advances in phytochemistry, Vol. 5: Structural and functional aspects of phytochemistry, pp. 197–224. Runeckles, V. C., Tso, T. C. (ed.). New York: Academic Press 1972
22. Baker, J. E., Norris, D. M.: J. Insect Physiol. *17*, 2383–2394 (1971)
23. Ma, W.-C.: Meded. Landhoogegsch., Wageningen *72*, 192 p. (1972)
24. Buhr, H., Toball, R., Schreiber, K.: Entomol. Exp. Appl. *1*, 209–224 (1958)
25. Geissman, T. A.: Flavonoid compounds, tannins, lignins and related compounds. In: Comprehensive biochemistry, Vol. *9*, pp. 213–250. Florkin, M., Stotz, E. A. (ed.). Amsterdam: Elsevier Publ. Co. 1963
26. Norris, D. M.: Repellents and deterrents in feeding of *Scolytus multistriatus*. In: Host plant resistance to pests, pp. 215–230. Hedin, P. (ed.) Washington DC: Am. Chem. Soc. 1977
27. Zanno, P. R. et al.: J. Am. Chem. Soc. *2* (1975)
28. Kubo, I., Miura, I., Nakanishi, K.: J. Am. Chem. Soc. *98*, 6704 (1976)
29. Munakata, K.: Insect antifeedants of *Spodoptera litura* in plants. In: Host plant resistance to pests, pp. 185–196. Hedin, P. A. (ed.). Washington, D. C.: Am. Chem. Soc. 1977
30. Mabry, T. J. et al.: Antifeedant sesquiterpene lactones in the Compositae. In: Host plant resistance to pests, pp. 179–184. Hedin, P. A. (ed.). Washington D. C.: Am. Chem. Soc. 1977
31. Sharma, R. N., Joshi, V. N.: Biovigyanam *3*, 225 (1977)
32. Stipanovic, R. D.: Function and chemistry of plant trichomes and glands in insect resistance. In: Plant resistance to insects, pp. 69–100. Hedin, P. A. (ed.). Washington D. C.: Am. Chem. Soc. 1983
33. Langenheim, J. H. et al.: Biochem. Syst. Ecol. *6*, 299 (1978)
34. Stipanovic, R. D., Bell, A. A., Lukefahr, M. J.: Natural insecticides in cotton (*Gossypium*). In: Host plant resistance to pests, pp. 197–214. Hedin, P. A. (ed.). Washington D. C.: Am. Chem. Soc. 1977
35. Norris, D. M.: Nature *222*, 1263–1264 (1969)
36. Norris, D. M.: Ann. Entomol. Soc. Amer. *63*, 476–478 (1970)
37. Rodriguez, E., Levin, D. A.: Rec. Adv. Phytochem. *10*, 214–270 (1976)
38. Lichtenstein, E. P., Strong, F. M., Morgan, D. G.: J. Agric. Food Chem. *10*, 30–33 (1962)

39. Ryan, C. A.: Insect-induced chemical signals regulating natural plant protection responses. In: Variable plants and herbivores in natural and managed systems, pp. 43–60. Denno, R. F., McClure, M. S. (ed.). New York: Academic Press 1983
40. Bishop, P. D. et al.: Proc. Natl. Acad. Sci. U.S.A. *78*, 3536–3540 (1981)
41. Hahn, M. G., Darvill, A. G., Albersheim, P.: Plant Physiol. *68*, 1161–1169 (1981)
42. Lee, S.-C., West, C. A.: Plant Physiol. *67*, 633–639 (1981)
43. Carroll, C. R., Hoffman, C. A.: Science *209*, 414–416 (1980)
44. Loper, G. M.: Crop Sci. *8*, 104–106 (1968)
45. Akazawa, T., Uritani, I., Kubota, H.: Arch. Biochem. Biophys. *88*, 150–156 (1960)
46. Kogan, M., Paxton, J.: Natural inducers of plant resistance to insects. In: Plant resistance to insects, pp. 153–171. Hedin, P. A. (ed.). Washington D. C.: Am. Chem. Soc. 1983
47. Hori, K., Atalay, R.: Appl. Ent. Zool. *15*, 234–241 (1980)
48. Hori, K.: Appl. Ent. Zool. *8*, 103–112 (1973)
49. Liener, I. E., Kakade, M. L.: Protease inhibitors. In: Toxic constitutents of plant foodstuffs, pp. 8–66. Liener, I. E. (ed.). New York: Academic Press 1969
50. Keen, N. T., Bruegger, B.: Phytoalexins and chemicals that elicit their production in plants. In: Host plant resistance to pests, pp. 1–26. Hedin, P. A. (ed.). Washington D. C.: Am. Chem. Soc. 1977
51. Miles, P. W.: Annu. Rev. Phytopathol. *6*, 137–164 (1968)
52. Norris, D. M.: How insects induce disease. In: Plant disease Vol. *4*: How pathogens induce disease, pp. 239–255. Horsfall, J. G., Cowling, E. B. (ed.). New York: Academic Press 1979
53. Galun, R.: Proc. Internat. Congr. Entomol. *15*, 257–265 (1977)
54. Duffey, S. S.: Proc. Internat. Congr. Entomol. *15*, 323–394
55. Blum, M. S.: Chemical defenses of arthropods. New York: Academic Press 1981
56. Norris, D. M.: Chemoreceptor proteins. In: Neurotoxicology of insecticides and pheromones, pp. 59–77. Narahashi, T. (ed.). New York: Plenum Publishing Corp. 1979
57. Norris, D. M.: Possible unifying principles in energy transduction in the chemical senses. In: Perception of behavioral chemicals, pp. 289–306. Norris, D. M. (ed.). Amsterdam: Elsevier/North-Holland Biomedical Press B. V. 1981
58. Norris, D. M. et al.: Science *170*, 754–755 (1970)
59. Rozental, J. M., Norris, D. M.: Nature *244*, 370–371 (1973)
60. Dethier, V. G.: The hungry fly. Cambridge, MA: Harvard University Press 1976
61. Dethier, V. G.: Quart. Rev. Biol. *30*, 348–371 (1955)
62. Beets, M. G. J.: Structural aspects of chemoreception. In: Sensory properties of foods, pp. 71–89. Birch, G. G., Brennan, J. C., Parker, K. J. (ed.). London: Applied Science 1977
63. Birch, G. G.: A model for gustatory transduction. In: Perception of behavioral chemicals, pp. 187–204. Norris, D. M. (ed.). Amsterdam: Elsevier/North-Holland Biomedical Press B. V. 1981
64. Norris, D. M. et al.: J. Insect Physiol. *17*, 85–97 (1971)
65. Norris, D. M. et al.: Comp. Biochem. Physiol. *57C*, 55–59 (1977)
66. Norris, D. M.: A molecular and submolecular mechanism of insect perception of certain chemical information in their environment. In: Comportement des insectes et milieu trophique, pp. 81–102. Labeyrie, V. (ed.). Paris: Centre National de la Recherche Scientifique 1977
67. Norris, D. M. et al.: Contrib. Boyce Thompson Inst. *24*, 263–274 (1970)
68. Rozental, J. M., Norris, D. M.: Life Sciences *17*, 105–110 (1975)
69. Hara, T. J.: Perception of behavior-altering chemicals in aquatic environments. In: Perception of behavioral chemicals, pp. 29–57. Norris, D. M. (ed.). Amsterdam: Elsevier/North-Holland Biomedical Press B. V. 1981
70. Vande Berg, J. S.: Ultrastructural and cytochemical parameters of chemical perception. In: Perception of behavioral chemicals, pp. 103–131. Norris, D. M. (ed.). Amsterdam: Elsevier/North-Holland Biomedical Press B. V. 1981
71. Ma, W.-C.: Receptor membrane function in olfaction and gustation: Implications from modification by reagents and drugs. In: Perception of behavioral chemicals, pp. 267–287. Norris, D. M. (ed.). Amsterdam: Elsevier/North-Holland Biomedical Press B. V. 1981
72. Borg, T. K., Norris, D. M.: Z. Zellforsch. *113*, 13–28 (1971)
73. Singer, G., Norris, D. M.: Comp. Biochem. Physiol. *46* **B**, 43–56 (1973)
74. Singer, G., Rozental, J. M., Norris, D. M.: Experientia *30*, 1384–1385 (1974)
75. Singer, G., Rozental, J. M., Norris, D. M.: Nature *256*, 222–223 (1975)

76. Perlstein, M. T., Atassi, M. Z., Cheng, S. H.: Biochem. Biophys. Acta *236*, 174–182 (1971)
77. Phillips, C. E., Vande Berg, J. S.: J. Insect Physiol. *22*, 425–429 (1976)
78. Wetlaufer, D. B., Ristow, S.: Annu. Rev. Biochem. *42*, 135–158 (1973)
79. Szent-Györgyi, A.: The development of bioenergetics. In: Membrane structure and mechanisms of biological energy transduction, pp. 1–4. Avery, J. (ed.). London: Plenum Press 1973
80. Mitchell, P.: Chemiosmotic coupling in energy transduction: A logical development of biochemical knowledge. In: Membrane structure and mechanisms of biological energy transduction, pp. 5–24. Avery, J. (ed.). London: Plenum Press 1973
81. Mitchell, P.: Science *206*, 1148–1159 (1979)
82. O'Connell, R. J.: The encoding of behaviorally important odorants by insect chemosensory receptor neurons. In: Perception of behavioral chemicals, pp. 133–163. Norris, D. M. (ed.) Amsterdam: Elsevier/North-Holland Biomedical Press B. V. 1981
83. Norris, D. M., Chu, H. M.: J. Insect Physiol. *20*, 1687–1696 (1974)
84. Baker, J. E., Norris, D. M.: J. Insect Physiol. *17*, 2383–2394 (1971)
85. Davis, E. E., Sokolove, P. G.: J. Comp. Physiol. *105*, 43–54 (1976)
86. Singer, G., Norris, D. M.: Experientia *29*, 1483–1484 (1973)
87. Norris, D. M. et al.: Interactions between repellent quinones and purified receptor lipoprotein from *Periplaneta americaa* sensory neurons. In: Olfaction and taste, pp. 68. Le Magnen, J., MacLeod, P. (ed.). London, Washington DC: Information Retrieval Ltd. 1977
88. Norris, D. M., Ryan, M. F., Piotrowski, J. J.: Trans. Amer. Soc. Neurochem. *8*, 191 (1977)
89. Norris, D. M.: Assoc. Chemoreception Sci. Abst. 2 (1980)
90. Norris, D. M.: Arch. Insect Biochem. Physiol. (submitted)
91. Norris, D. M., Cary, L.: Insect Biochem. *11*, 743–750 (1981)
92. Schwartz, A., Lindenmayer, G. E., Allen, J. C.: Pharmacol. Rev. *27*, 3–134 (1975)
93. Bonting, S. L.: Sodium-potassium activated adenosinetriphosphatase and cation transport. In: Membranes and ion transport, pp. 257–363. Bittar, E. E. (ed.). London: John Wiley and Sons 1970
94. Anstee, J. H., Bell, D. M.: Insect Biochem. *8*, 3–9 (1978)
95. Grasso, A.: Life Sci. *6*, 1911–1918 (1967)
96. Schin, K., Kroeger, H.: Insect Biochem. *10*, 113–117 (1980)
97. Tolman, J. H., Steele, J. E.: Insect Biochem. *6*, 513–517 (1976)
98. Emmelot, P., Bos, C. J.: Biochem. Biophys. Acta *120*, 369–382 (1966)
99. Gervais, A. et al.: Circ. Res. *40*, 8–14 (1977)
100. Thomas, R. C.: Physiol. Rev. *52*, 563–594 (1972)
101. Skou, J. C.: Quart. Rev. Biophys. *7*, 401–434 (1975)
102. Cary, L. R.: M. S. Thesis. Madison: University of Wisconsin. 96p. 1980
103. Perkel, D. H., Bullock, T. H.: NRP Bull. *6*, 221–348 (1968)
104. Villet, R. H.: Nature *248*, 707–709 (1974)
105. Webb, J. L.: Enzyme and metabolic inhibitors III. New York: Academic Press 1966
106. Friedman, M.: The chemistry and biochemistry of the sulfhydryl group in amino acids, peptides and proteins. Oxford: Pergamon Press 1973
107. Norris, D. M.: Symp. Biol. Hung. *16*, 197–201 (1976)
108. Noma, A., Hiji, Y.: Jap. J. Physiol. *22*, 393–402 (1972)
109. Shimada, I., et al.: J. Insect Physiol. *18*, 1845–1855 (1972)
110. Shimada, I.: J. Insect Physiol. *21*, 1665–1674 (1975)
111. Kurihara, K., Kamo, N., Kobatake, Y.: Adv. Biophys. *10*, 27–95 (1978)
112. Kashiguwara, T., et al.: Brain Res. *142*, 570–575 (1978)
113. Naráhashi, T.: Physiol. Rev. *54*, 813–889 (1974)
114. Akaike, M., Sato, M.: Jap. J. Physiol. *26*, 29–40 (1976)

Subject Index

Chemie der Pflanzenschutz- und Schädlingsbekämpfungsmittel

Herausgeber: **R. Wegler**

Band 8

Spezielle Chemie der Herbizide Anwendung und Wirkungsweise Special Chemistry of Herbicides Applications and Mechanisms

1982. VII, 455 Seiten (40 Seiten in Englisch). ISBN 3-540-10778-9

Contents/Information: Seit im Jahr 1977 Band 5 des Werkes zum Thema „Herbizide" erschien, gewannen die Herbizide erneut an Bedeutung. Neben der Einführung von Versuchs-und Handelsprodukten, die Substanzgruppen entstammen, deren Bedeutung schon Ende 1976 bekannt war, wurden auch spektakuläre Entdeckungen gemacht. Herausgeber und Verlag haben sich daher entschlossen, diesen Band den Neuentwicklungen auf dem Gebiet der Herbizide zu widmen.

Schon 1976 bekannte Verbindlungsklassen, wie die der Diarylether-oxyalkancarbonsäuren, ergaben eine Fülle neuer Herbizide mit meist selektiver Gräserwirkung im Getreide. Mitarbeiter der Firma Hoechst AG haben daher für diesen Band einen entsprechenden Beitrag zur Verfügung gestellt.

H. J. Nestler: **Phenoxy-phenoxy-Propionsäure-Derivate und verwandte Verbindungen (Phenoxy-phenoxy propionic Acid Derivatives and Related Compounds)**

Das immer noch aktuelle Problem der Wildhaferbekämpfung wird in einem Beitrag der Shell Int. Res. behandelt, eine Firma mit speziellen Erfolgen bei Flughafer-Herbiziden.

E. Haddock, R. G. Turner: **Wild Oat Herbicides (Wildhafer-Herbizide)**

Der Herausgeber selbst hat die Herbizid-Fortschritte 1976 bis 1980 bearbeitet, einschließlich einer vollständigen Patentübersicht.

R. Wegler, L. Eue: **Neue Herbizide (New Herbicides)**

Erstmals werden neben Patenten bzw. Offenlegungsschriften der Bundesrepublik Deutschland die an Bedeutung rasch zunehmenden Europa-Patente vollständig aufgeführt. In Anbetracht der wachsenden Bedeutung Japans als Erfinder, Hersteller und Verbraucher von Pflanzenschutzmitteln, wurden japanische Patente mit besonderer Sorgfalt erfaßt. Zusammen mit der Patentergänzung aus Band 5 liegt nun ein fast lückenloser Überblick über die wichtigsten Patente vor. Aus den Patentergänzungen lassen sich wichtige Hinweise auf Neuentwicklungen ablesen, was für eine ökonomische Forschungsplanung von großer Bedeutung ist.

Springer-Verlag
Berlin
Heidelberg
New York
Tokyo

Chemie der Pflanzenschutz- und Schädlingsbekämpfungsmittel

Herausgeber: **R. Wegler**

Band 6

Insektizide. Bakterizide. Oomyceten-Fungizide. Biochemische und biologische Methoden. Naturstoffe Insecticides. Bactericides. Oomycete Fungicides. Biochemical and Biological Methods. Natural Products.

1981. 105 figures, 92 schemes. XVI, 512 pages. (191 pages in German). ISBN 3-540-10307-4

Contents:

Springer-Verlag Berlin Heidelberg New York Tokyo

conjugated dienes.